A DICTIONARY OF SPECTROSCOPY

Second edition

A DICTIONARY OF SPECTROSCOPY

SECOND EDITION

R.C. DENNEY

Macmillan Reference Books

First published 1982 by
THE MACMILLAN PRESS LTD
London and Basingstoke
Associated Companies throughout the world

ISBN 0 333 31670 3

Typeset by Leaper & Gard Ltd, Bristol
Printed in Great Britain
by Billing and Sons Limited
Guildford, London, Oxford, Worcester

Yes, I have a pair of eyes', replied Sam, 'and that's just it. If they was a pair o' patent double million magnifyin' gas microscopes of hextra power, p'raps I might be able to see through a flight o' stairs and a deal door; but bein' only eyes, you see my wision's limited.'

Charles Dickens
Pickwick Papers, Ch. 34

Acknowledgements

During the preparation of the second edition of this book I have again become aware of the encouragement and assistance of many friends and colleagues who have once again provided comments, suggestions and diagrams for the new work. I am particularly grateful to my good friend Dr Ralph Thomas who has given freely of his time to advise me on the entries dealing with photoelectron spectroscopy. I must, however, accept full responsibility for any errors or omissions in these or any other parts of the new edition.

I have been greatly encouraged throughout the rewriting by the enthusiasm of my publishers who have been anxious to build upon the earlier success. A number of the diagrams have been kindly provided by instrument manufacturers or permission granted by other authors and I greatly appreciate their kindness in this, they are:

Figure 8,   Perkin-Elmer Ltd, Beaconsfield, Buckinghamshire, UK

Figure 12,   Rank Precision Industries Ltd (Hilger and Watts), Analytical Division)

Figure 14,   AEI Scientific Apparatus Ltd

Figure 27,   Perkin-Elmer Ltd

Figure 29,   Specac, Unit 3, Lagoon Rd, St Mary Cray, Kent (drawn by Twyprint, 19 Cranleigh Drive, Swanley, Kent)

Figures 31 and 48, Prof. W.R. Brode from *Chemical Spectroscopy*, 2 edn, J. Wiley (1942)

Figure 33,   Dr. W. McFarlane from *Chemistry in Britain*.

Figure 44,   F.W. Fifield and D. Kealey, from *Principles and Practice of Analytical Chemistry*, International Textbook Co. Ltd (1975).

All other diagrams are the copyright of Dr. R.C. Denney and/or have been redrawn from the first edition of this book.

Once again I owe a great debt to my wife and children who have accepted the curtailment of my free time in an effort to produce this and its companion volume *A Dictionary of Chromatography* in a limited period of time in order that they could appear together.

Introduction to the Second Edition

During the nine years since the first edition of this book was produced it has become even clearer that the need for a relatively simple reference book dealing with spectroscopic terms exists both amongst students and people who are not generally familiar with developments in spectroscopy. I hope that by collecting together the most common expressions, equations and terms in an alphabetical form I will have provided a useful reference book as well as a source to other papers, articles and reviews dealing with the various subject areas. The book is not intended to be a detailed treatise on each topic suitable for specialists in atomic or molecular spectroscopy, for that it would need to be much larger and more expensive.

Throughout the book I have tried to standardize and up-date the symbols and nomenclature in line with the most recent recommendations, although some variations in terminology still persist. As far as possible I have adhered to those published by the Symbols Committee of the Royal Society[1] and by the International Union of Pure and Applied Chemistry[2] employing SI units.

This edition is substantially larger than the previous one mainly because photoelectron spectroscopy has been included and the number of diagrams increased, despite this the entries have been kept as concise as is practicable. I have incorporated a number of new entries following suggestions from reviewers and colleagues and hope that most of the previous gaps have now been filled.

I have continued the previous practice of following many headings with an abbreviation in parentheses. This indicates the main, but not necessarily exclusive, realm of spectroscopy to which it applies. The following abbrevations have been employed for this purpose:

(a.s.)	atomic spectroscopy
(em.)	emission
(e.s.r.)	electron spin resonance
(i.r.)	infrared
(m.s.)	mass spectrometry
(n.m.r.)	nuclear magnetic resonance
(pe.)	photoelectron
(pl.)	photoluminescence
(r.s.)	Raman spectroscopy
(u.v.)	ultraviolet/visible

I hope many readers will find the book of value in their work and would be pleased to receive any constructive suggestions for improvement of or for further inclusion in a future edition.

Ronald C. Denney

Symbols

The following symbols and letters are employed in the equations throughout the book. Symbols used only once are explained in the text.

A	ampere	I	Intensity of transmitted light
A	absorbance		
Å	ångström	I	spin quantum number
A_r	atomic mass of an element	I_0	intensity of incident light
a	absorption coefficient	J	joule
a_i	coupling constant	J	coupling constant
B	magnetic flux density	K	kelvin
c	concentration	k	Boltzmann constant
c	speed of light in a vacuum	k	kilo
cm	centimetre	kg	kilogram
E	energy	L	litre
E_F	Fermi energy	L	Avogadro constant
e	base of natural logarithms	l	cell path length
e	elementary charge	l	time-of-flight path length
f	force constant	M	mass of nucleus
g	gram	M	radiant exitance
g	Landé factor	M	magnetization
H	henry (inductance)	m	metre
H	magnetic field strength	m	mass
H_1	radio-frequency field amplitude	m_0	mass of electron at rest
		m_1	magnetic quantum number
h	Planck constant	m_n	mass of neutron at rest

m_p	mass of proton at rest	α	polarizability
mm	millimetre	γ	magnetogyric ratio
N	number of molecules in the ground state	δ	chemical shift
		ϵ	molar absorption coefficient
N_A	Avogadro constant	θ	angular measurement
N_E	number of molecules with energy E	λ	wavelength
		μ	magnetic moment of particle
n	refractive index		
n	spectrum order	μ	permeability
nm	namometre	μ	statistical mean
P	probability	μ	micro
Pa	pascal	μ_B	Bohr magneton
Q	quadrupole moment	μ_N	nuclear magneton
R	gas constant	μ_0	permeability of vacuum
R_H	Rydberg constant for hydrogen	μ_1	relative permeability
		ν	frequency
R_∞	Rydberg constant (heavy atom)	$\bar{\nu}$	wavenumber
		π	ratio of circumference to diameter of circle
r	path radius		
r_S	distance from scattering particle	ρ_N	depolarization factor
		σ	nuclear screening constant
T	tesla	σ	standard deviation
T	temperature	σ	Stefan-Boltzmann constant
T	transmittance	τ	chemical shift
T_1	longitudinal relaxation time	Φ	magnetic flux
T_2	transverse relaxation time	χ_m	magnetic susceptibility
T_S	spin temperature	Ψ	eigenfunction (work function)
t	time		
V	volt	ω_L	Larmor (angular) frequency
v	velocity	ω_N	nuclear angular precession frequency
W	watt		
Z	charge on nucleus		

Greek Alphabet

As so many spectroscopic equations include the use of Greek letters the following alphabet has been included to help the reader to give the correct name to these letters as they arise in the text:

alpha	A	α	nu	N	ν
beta	B	β	xi	Ξ	ξ
gamma	Γ	γ	omicron	O	o
delta	Δ	δ	pi	Π	π
episilon	E	ϵ	rho	P	ρ
zeta	Z	ζ	sigma	Σ	σ
eta	H	η	tau	T	τ
theta	Θ	θ	upsilon	Υ	υ
iota	I	ι	phi	Φ	ϕ
kappa	K	κ	chi	X	χ
lambda	Λ	λ	psi	Ψ	ψ
mu	M	μ	omega	Ω	ω

A

AAS. *See* **Atomic absorption spectroscopy.**

A bands (u.v.). Various systems of nomenclature for ultraviolet spectra have designed different absorption groups as A bands. Moser and Kohlenberg[4] used it to refer to the bands occurring at the shortest-wavelength region. *See* **Bands (ultraviolet).**

Abscissa. The name given to the horizontal axis in a set of two-dimensional coordinates. For convenience it is commonly referred to as the X axis, so that the abscissa of any point in the plane of the two axes forming the coordinates is the value at the point on the X axis at which a vertical line from the point meets that axis. *See also* **Ordinate.**

Absorbance (A). The absorbance for a material is the logarithm of the ratio of the intensities of the incident light and the transmitted light. It is a dimensionless quantity the value of which can be calculated from the **molar absorption coefficient**, the molar concentration and the path length through the material according to the equation:

$$A = \log_{10} \frac{I_0}{I} = \epsilon c l = \log_{10} \frac{I}{T}$$

The term absorbance has been recommended for use in preference to absorbancy, optical density or extinction.[5] *See also* **Beer–Lambert law.**

Absorbancy. *See* **Absorbance.**

Absorption. Transitions from low energy states to higher energy states within molecular and atomic systems occur as a result of absorption of electromagnetic radiation. Changes in electronic energy levels are measured by absorption of ultraviolet/visible radiation; changes in vibrational energy levels occur by absorption of infrared radiation and

1

changes in nuclear energy levels by absorption of radio-frequency radiation.

Similarly the absorption of **microwave radiation** is used to study transitions between electron spin energy levels in **electron spin resonance** spectroscopy.

Absorption band. Any region of the absorption spectrum in which the **absorbance** includes a maximum. *See also* **Bands (infrared)** and **Bands (ultraviolet)**.

Absorption coefficient (a). *See* **Extinction coefficient**.

Absorption spectrum. A record of the **absorbance** or **transmittance** of a material with respect to wavelength or some function of the wavelength which may be plotted manually or automatically recorded.

Absorptivity. *See* **Extinction coefficient**.

Absorptivity (M). *See* **Molar absorption coefficient**.

Abundance tables (m.s.). Lists of molecular formulae, usually limited to compounds containing carbon, hydrogen, oxygen and nitrogen, showing the relative sizes of the $M + 1$ and $M + 2$ peaks to the **parent peak** for each formula. These particular peaks arise from the existence of the heavier isotopes in the molecules and their sizes are related to the number of each atomic species present. These tables[6] are now based upon $^{12}C = 12.0000$; earlier tables were based upon $^{16}O = 16.0000$. *See also* **Isotopic abundance**.

a.c. arc source (em.). Initial voltages up to 5000V are necessary to achieve ignition of the arc in this method of vaporization and excitation in emission spectroscopy, but the operating level is then maintained at about 50V. The arc is struck between an electrode gap of about 2.5 mm with a current of 2 to 5 A. Results are more reproducible than with the d.c. arc. The a.c. arc source is considered a compromise between the poor sensitivity, high selectivity of the **a.c. spark source** and the high sensitivity, poor selectivity of the **d.c. arc source**. It is used particularly with zinc and other non-ferrous metal alloys, although it tends to give a greater cyanogen background from the atmosphere than does the a.c. spark source.

Accelerating slits (m.s.). Positively charged ions produced in the ionization chamber of a mass spectrometer are passed into the analyser of the instrument by the accelerating slits. For this purpose two accelerating slits are employed. The first is at a slight negative potential with respect to the ionization chamber in order to attract the positive ions from the source. The real acceleration occurs as a result of the powerful electrostatic field of up to 8000 V between the first and second slits which leads to ion velocities of up to 150 000 m s^{-1} being produced. As a result the ions possess identical kinetic energies but different velocities; as given by the equation

$$e\mathrm{V} = \tfrac{1}{2}m_1 v_1{}^2 = \tfrac{1}{2}m_2 v_2{}^2 = \tfrac{1}{2}m_3 v_3{}^2, \text{etc}$$

Achromatic lens. A transparent substance is said to be achromatic when it has the ability of transmitting light without dispersing it into its constituent colours. Thus, an achromatic lens will produce images which do not have rainbow–coloured edges. This is achieved by forming the lens from two materials possessing different refractive indices such that the **dispersion** due to one material is corrected by the dispersion due to the other, in practice the correction is only absolute at two selected wavelengths, some slight dispersion remaining for the other wavelengths in the spectrum. Crown glass and flint glass are the most common combination for this purpose.

a.c. spark source (em.). The preferred type of source for high-precision emission spectroscopy, particularly for producing ionic spectra. The spark is produced from a transformer, capable of operating as high as 50 kV, connected in parallel with a condenser across the spark gap. The heating effect of the spark source is less than that of either the **a.c. arc source** or the **d.c. arc source** and is used for studying low-melting solids and liquids. Graphite electrodes are usually used and the spark energy

4 *Adiabetic ionization energy*

provides the excitation energy necessary for electronic transitions to take place. Temperatures as high as 10000 K may be attained during the process.

Adiabetic ionization energy (m.s.). *See* **Ionization energy.**

AFS. *See* **Atomic fluorescence spectroscopy.**

Analyser (pe.). *See* **Electron energy analyser.**

Analyser tube (m.s.). In a mass spectrometer the section of the instrument in which positive ions are separated according to their mass/charge ratios. The operating pressure for the tube is about 10^{-5} to 10^{-6} Pa (10^{-7} to 10^{-8} torr). In mass spectrometers using electrostatic and magnetic fields the analyser tube is curved and almost completely enclosed by the fields; in the **quadrupole mass spectrometer** and the **time-of-flight mass spectrometer** the tube is straight. *See also* **Field-free region.**

Ångstrom Å. The unit of length equal to 1/6438.4696 of the wavelength of the red line of cadmium.

$$1 \text{ Å} = 10^{-10} \text{ m} = 10^{-8} \text{ cm}$$

The angström has been a useful dimension for representing interatomic distances; and continues to be used for this purpose although it is not an **SI unit.**

Anisotopic (m.s.). An element is said to be anisotopic when it possesses only one naturally occurring isotope. Fluorine is a typical anisotopic element and mass spectra of molecules containing fluorine atoms are more simple than those of corresponding chlorine and bromine compounds.

Anisotropy (n.m.r.). *See* **Diagrammatic anisotropy.**

Anode (m.s., pe.). The positive electrode in an electron tube or electron ionization chamber which acts as a collector for the electrons and any negatively charged particles. It therefore operates at a positive potential with respect to the **cathode**. In the ionization chambers of mass spectrometers the anode is known as the 'trap' and is made from a robust metal such as stainless steel.

For photoelectron spectroscopy the generation of X-rays is achieved by bombarding water cooled anodes with high energy electrons from an **electron gun**. The anodes (or targets) employed are most commonly aluminium or magnesium coatings on a copper base. These can operate with surface temperatures up to 400 °C and with anode potentials below 15 kV to produce characteristic $K_{\alpha 1,2}$ emission lines with narrow bandwidths.[7] In some cases twin anode sources (both Al and Mg targets) are used.

Anti-bonding orbitals (u.v.). *See* **Bonding orbitals**.

Anti-Stokes fluorescence (pl.). Although most fluorescent species comply with the **Stokes law**, a weak fluorescence found at a shorter wavelength than the exciting wavelength is also sometimes observed at room temperature. This anti-Stokes fluorescence derives its additional energy from excited vibrational levels within the ground state.[8]

Anti-Stokes lines (r.s.). The Raman lines measured on the shorter-wavelength side of the monochromatic radiation source. They arise from those Raman transitions in which the final vibration level is lower than the initial vibration level. The anti-Stokes lines in Raman spectroscopy are much weaker than are the corresponding **Stokes lines** occurring at longer wavelengths; for this reason the latter spectrum is preferentially scanned. *See* **Raman spectroscopy**.

Apparent mass (m.s.). *See* **Metastable ions**.

Appearance potential (m.s.). Fragmentation ions in a mass spectrum do not occur until the **ionization energy** for the molecule has been exceeded. In general the appearance potential for the most readily formed fragment ions is about 1 to 4 eV higher than the value for the ionization energy of the corresponding molecular ion.

Arc source (em.). *See* **a.c. arc source** and **d.c. arc source.**

Astigmatic. Optical systems in spectrometers are said to be astigmatic when the focal point for the vertical image occurs at a different position from that for the horizontal image. This is a problem that occurs with many forms of diffraction grating mounts.

Atomic absorption spectroscopy (AAS) (a.s.). A very sensitive method of spectroscopy for elemental analysis in which the sample is converted into an atomic vapour. The process produces pure electronic transitions free from vibrational and rotational transitions. The number of atoms capable of absorbing any particular wavelength of transmitted light is proportional to the concentration of these atoms in the flame and to the length of the path in the flame. In its simplest form, illustrated in Figure 1, the radiant source employed is a hollow **cathode** lamp emitting radiation from excited atoms of the same element as that to be determined. This radiation, in the form of a line spectrum, is absorbed by the sample which has been atomized at 2000–3000 °C either in a flame or an **electrothermal atomizer.** The resulting spectrum of the source radiation is one with lines of reduced intensity, the magnitude of the reduction being proportional to the concentration of the element in the atomized vapour. The monochromator in the instrument is used to select a single wavelength for examination and changes in the intensity of this wavelength are related to different concentrations of the nebulized sample.[9]

Atomic absorption has expanded very rapidly in a relatively short period[10,11] and its applications have been greatly extended by the use of flameless devices such as carbon rods and furnaces, and by the development of high temperature plasma atomization. As a result, detection limits as low as $10^{-2}–10^{-3}$ μg ml^{-1} can be attained with a precision better than 1 per cent.[12] *See also* **Electrodeless discharge tube.**

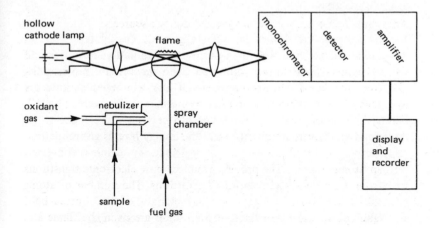

Figure 1. Layout of atomic absorption
spectrometer

Atomic fluorescence spectroscopy (AFS). The development of atomic fluorescence spectroscopy owes a great deal to that of **atomic absorption spectroscopy** although in this case it is a measurement of emitted radiation rather than absorbed radiation. In AFS the electrons in free atoms are promoted to excited electronic states but are identified by the characteristic fluorescent radiation lines emitted in all directions as the energy of the electrons drop to lower levels. By operating at a fixed, known fluorescent wavelength the process may be used for accurate quantitative determinations. The instrumentation required is essentially the same as for **AAS**, but has the advantage that, as the emitted radiation is measured out of direct line of the radiation source (usually at right angles), narrow line sources such as **hollow cathode lamps** are not essential and a **xenon arc lamp** is frequently adequate.[13]

AFS has the great advantage that it is possible to determine several elements simultaneously. It is, however, susceptible to effects arising from **quenching** and detection limits are frequently lower than those for AAS. Laser excited atomic fluorescence flame spectrometry has been shown[14] to be capable of detecting a wide range of elements at levels below 30 ng mL^{-1}.

Atomic spectra. Transitions between electronic energy levels within atoms lead to the emission or absorption of radiation in a series of sharply defined lines corresponding to fixed wavelengths representing radiation quanta of definite energies. Of the five series of lines for hydrogen that in the visible region was the first discovered, by J.J. Balmer.[15] The formula for the relationship between wavelength of the radiation (λ), the wavenumber ($\tilde{\nu}$) and the energy level is given by:

$$\frac{1}{\lambda} = \tilde{\nu} = R_H \left(\frac{1}{n_1^2} - \frac{1}{n_2^2} \right)^2$$

The values of n_1 and n_2 for the hydrogen series are given by Table 1.

Table 1

Series	n_1	n_2	Region
Lyman	1	2, 3, 4, . . .	far u.v.
Balmer	2	3, 4, 5, . . .	visible
Paschen	3	4, 5, 6, . . .	far i.r.
Brackett	4	5, 6, 7, . . .	far i.r.
Pfund	5	6, 7, 8, . . .	far i.r.

The general formula for hydrogen like spectra (produced from ions such as He$^+$, Li^{2+}, Be^{3+}) is

$$\tilde{\nu} = RZ^2 \left(\frac{1}{n_1^2} - \frac{1}{n_2^2} \right)$$

The value for the Rydberg constant in each case is determined from R_∞.

Atomizer (a.s.). In spectroscopy the word used specifically for that part of the instrument in which free atoms are released. In **atomic absorption spectroscopy** this refers to the flame or **electrothermal atomizers**, whereas the formation of liquid droplets is carried out by **nebulizers**.

Atom reservoirs (a.s.). *See* **Carbon filament atom reservoir**; **Delves cup**; **Electrothermal atomizers**; **Massman furnace**; **Tantalum boat**.

ATR. *See* **Attenuated total reflectance**.

Attenuated total reflectance (ATR) (i.r.). As infrared spectroscopic technique used particularly for examining opaque solids, in which the sample is placed against the face of a prism.[16] Radiant energy passing through the wall of the prism is selectively reflected by the sample, having only penetrated it by a few micrometres. The refractive index of the sample must be less than that of the prism and a good optical contact between the sample and prism must be maintained. For attenuated total reflectance to occur it is necessary to use a fixed angle of incidence that is greater than the critical angle. The prism (or hemicylinder) may be silver chloride, silicon or **KRS-5**[17] as these possess refractive indices > 2. In modern ATR systems multiple absorption of the light beam is achieved by using a polished trapezoid with a beam directed through the prism face at greater than the critical angle[18] (Figure 2). As a result, much stronger absorption spectra are obtained. The spectra produced by this procedure are similar to ordinary absorption spectra and can be obtained with micro quantities.[19]

Attenuation (e.s.r.). Attenuation of the microwave power passing along a **waveguide** is achieved by means of a metal plate placed along the axis of the waveguide. The degree of attenuation increases as the plate is moved away from the wall of the waveguide towards the centre.

Attenuator (i.r., a.v.). A toothed comb, grid or star wheel arrangement introduced into one beam of a spectrophotometer, operated either automatically through an electronic servosystem, or manually to balance the radiation in both beams. *See also* **Optical null principle**.

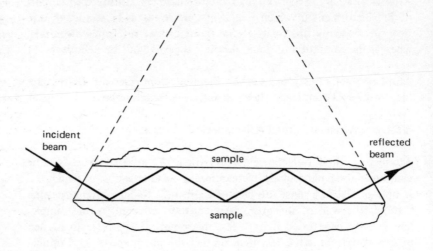

Figure 2. ATR using multiple absorption

Auer burner (i.r.). One of the sources of radiation suitable for far-infrared spectroscopy. The radiation is emitted from a thorium oxide mantle heated to 1800 °C by burning gas. Its optimum operating region is around 50 μm, above this a quartz mercury lamp is used.

Auger effect (pe.). The name given to secondary electron emission following reorganization that can occur within ions that have been formed by loss of an inner shell electron from neutral atoms.[20] The reorganization takes place by a higher-level electron filling the inner shell vacancy and in so doing transferring its energy to another electron which is ejected. This ejected electron is known as the Auger electron and the path followed may be studied by cloud-chamber photographs. Electron energy measurements, forming the basis of Auger electron spectroscopy (AES) are carried out using electrostatic energy analysers coupled to electron multipliers and a recorder.[21] The process is shown diagramatically in Figure 3.

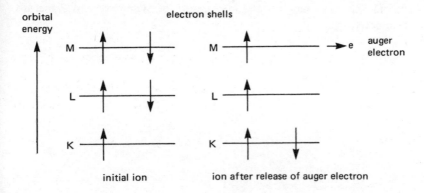

orbital
energy

electron shells

M

L

K

initial ion

M

L

K

e auger
electron

ion after release of auger electron

Figure 3. Auger process

Auxochromes (u.v.). Certain chemical groups substituted into molecules have been found to cause a change in the wavelength and magnitude of the characteristic absorptions arising from **chromophores**. These groups are called auxochromes and do not themselves usually absorb above 200 nm. Typical auxochromes are found to possess **non-bonding electrons** such as $-OH$, $-NH_2$ and $-SH$. It is considered that the auxochrome functions by a charge transfer mechanism,[22] the auxochrome acting as a donor of charge during the electronic transition of electrons in an adjacent bond. *See also* **Bathochromic shift** and **Hyperchromic effect**.

Avogadro constant (L or N_A). The number of molecules in 1 mole of any substance, currently given as $6.022\ 045 \times 10^{23}$ mole^{-1}.

Axis of symmetry (i.r.). When a molecule can be rotated about any axis passing through it to produce a molecular arrangement indistinguishable from the original position it is said to possess an axis of symmetry. If such a situation occurs for a rotation of $360°/2 = 180°$ it is said to be a two-fold axis of symmetry, and if for every $360°/3 = 120°$ it is a three-fold axis of symmetry. Linear molecules possess an ∞-fold axis of symmetry. *See also* **Centre of symmetry** and **Plane of symmetry**.

B

Balmer series (em.). *See* **Atomic spectra.**

Bandpass filter. In spectrophotometers incorporating diffraction gratings it is necessary to ensure that only first-order spectra are obtained. The bandpass filter is incorporated into the optical system to remove the higher-order spectra. It consists of two filters, one giving an increase in transmission from longer to shorter wavelength, the other giving the increase in transmission from shorter to longer wavelength. The useful region is, therefore, restricted to that portion in which the transmission ranges of the two filters overlap.

Bandpass width. *See* **Effective bandwidth.**

Bands (infrared). Spectra in the middle-infrared region of absorptions arising from vibrational transitions in molecules accompanied by rotational transitions. Complete resolution of the fine structure due to individual rotational transitions in this region can be carried out for gaseous samples. For liquid or solid samples infrared bands have the broad outline of the vibrational transition envelope. *See also* **Infrared spectra.**

Bands (ultraviolet). Ultraviolet/visible absorption spectra of conjugated organic compounds are considered to arise from four main electronic transition types giving rise to broad absorption bands. In order to establish some relationship between absorption bands from different compounds several classifications have been made, as shown in Table 2. Silverstein et al.[23] have shown the connection between some of these classification and related them to the typical transitions of each group.

Band spectra (em.). *See* **Emission spectrometry.**

Bandwidth. *See* **Effective bandwidth.**

13

Table 2. Absorption bands[24]

	increasing wavelength ⟶			
Burawoy[25] and Braude[26]	—	K and E	B	R
Doub and Vandenbelt[27]	2nd primary band	3rd primary band	secondary band	—
Klevens and Platt[28]	A and B	C	D	—
Moser and Kohlenberg[29]	A	B	C	D
Clar[30]	β	p	α	—
Type of transition	$\sigma \rightarrow \sigma^*$ and $n \rightarrow \sigma^*$	$\pi \rightarrow \pi^*$	$\pi \rightarrow \pi^*$	$\pi \rightarrow \pi^*$

Bar. *See* **Pressure, units.**

Bar graph (m.s.). A diagrammatic presentation of a mass spectrum using a linear mass/charge scale. In bar graphs the various spectral peaks are represented by vertical lines with lengths corresponding to the relative abundancies of the individual ions expressed as percentages with reference to the most abundant ion, as shown in Figure 4 (*see* **Base peak**).

Barrier-layer cell (u.v., a.s.). A photoelectric detector system consisting of a thin film of silver covering a layer of semiconductor, usually selenium, deposited upon an iron base. After passing through the silver film the radiation falling on the semiconductor leads to the release of electrons at the selenium–silver interface and these pass to the silver layer which acts as a collector electrode. The area of contact between the conducting metal and semiconductor is the barrier layer.

Base-line method (i.r.). In quantitative analysis one of the major difficulties in measuring absorbance values is the very irregular base line. To overcome this problem an arbitrary line may be chosen to represent the base line for the **absorption band** being employed for the measurement.[31] In practice the base line is usually drawn as a tangent to two adjacent absorption maxima and the vertical distance from this line to the absorption maximum is the measurement employed in the quantitative calculation. The method has been used for the quantitative analysis

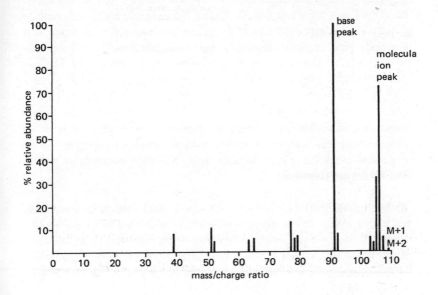

Figure 4. Bar graph for mass spectrum of p-xylene

of mixtures[32] by infrared spectroscopy and may also be applied in other realms of spectroscopy.

Base peak (m.s.). The most intense peak in a mass spectrum. It is given the arbitrary value of 100 in order that all other peaks may be expressed as a percentage of the base peak (*see* **Bar graph**). The **parent** (or **molecular**) **peak** is not necessarily the base peak as in most mass spectra it is not the most intense peak obtained. Frequently the base peak is that obtained for a fragment ion formed by loss of a small portion of the parent molecule.

Batch inlet sampling system (m.s.). The purpose of the batch inlet system for a mass spectrometer is to enable a volatile sample to be continuously fed into the ionization chamber without having to add repeated small amounts of sample. For this purpose the main sample, contained in a tube that can be cooled in liquid nitrogen, is allowed to expand into a large reservoir of up to one litre capacity. From here it is gradually bled through a leak valve (*see* **Molecular leak**) into the ionization chamber. The reservoir is maintained at a pressure of 10^{-5} to 10^{-6} torr whilst the mass spectrometer is kept at about 10^{-7} to 10^{-8} torr. By this means the spectra of gases and volatile liquids can be obtained without the risk of either spoiling the vacuum in the instrument or overloading the ionization chamber. Once the material from the reservoir has been exhausted a further amount of sample can be expanded from the cooled sample tube. *See also* **Direct inlet system** and **Direct insertion probe.**

Bathochromic shift (u.v.). When structural modification in a molecule leads to a characteristic absorption maximum shifted to a longer wavelength (lower frequency) it is said to have caused a bathochromic shift. This is sometimes referred to as a red shift, as the displacement is in the direction of the red end of the visible spectrum. *See also* **Hypsochromic shift.**

B bands (u.v.). Under the Moser and Kohlenberg[33] nomenclature the B bands are those corresponding to the most powerful electronic transitions giving rise to strong absorptions, usually above 220 nm with **molar absorption coefficients** exceeding 5000. Braude[34] used the letter B to indicate absorptions arising from typical benzenoid structures. *See also* **Bands (ultraviolet).**

Beer, Bernard, Bouguer and Lambert laws. Considerable confusion exists over the prior claim and mode of reference to these individual laws and their combined form.

The fundamental law relating to absorption and thickness of the medium is frequently credited to Lambert,[35] although he actually restated and extended a law that had been originally enunciated by Bouguer.[36] The second fundamental law dealing with absorption and its relation to conconcentration was published by Beer[37] shortly prior to its publication in the same year by Bernard.[38] The whole of this very confused story has been sorted out by Malinin and Yoe.[39]

Generally the combined form of the two related laws is referred to as the **Beer–Lambert law**.

Beer–Bouguer law (i.r., u.v.). *See* **Beer–Lambert law**.

Beer–Lambert law (i.r., u.v.). The combined form of the absorption laws is obtained[40] by integrating the equation

$$dI = \left(\frac{\partial I}{\partial l}\right)_c dl + \left(\frac{\partial I}{\partial c}\right)_l dc$$

giving a mathematical expression of the form

$$\log \frac{I_0}{I} = acl$$

in which a is a constant and l and c are empirical units for sample path length and concentration respectively.

The more specific form of this equation is

$$\log \frac{I_0}{I} = \epsilon cl.$$

In this case the constant ϵ is the **molar absorption coefficient**, c is molar concentration and l is the path length (usually in cm). *See also* **Absorbance**.

Beer law (u.v., i.r.). The portion of incident radiation energy absorbed by a material in solution is proportional to the concentration of that material. Mathematically this is expressed as

$$-\frac{dI}{dc} \propto I$$

which on integration becomes

$$\log \frac{I_0}{I} = k_2 c$$

See also **Absorbance** and **Beer–Lambert laws.**

Bending vibrations (i.r.). For a non-linear molecule consisting of n atoms there are $3n-6$ vibrational modes ($3n-5$ for a linear molecule), and of these $2n-5$ are bending (or deformation) vibrations. In these vibrations the motion occurs in a direction perpendicular to the bond between the atoms. Bending vibrations are frequently subdivided into four types: rocking, scissor, twisting and wagging. *See also* **Stretching vibrations.**

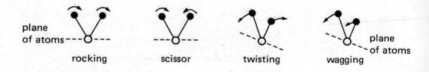

plane
of atoms

rocking   scissor   twisting   wagging

plane
of atoms

Figure 5. Bending vibrations

Bernard law. *See* **Beer law.**

Binding energy (pe.). Apart from its connection in nuclear energy, the term binding energy is also used in photoelectron spectroscopy to refer to the energy necessary to bring about the ionization of an atom by loss of a particular electron. Thus in the equation

$$h\nu = I_k + E_k$$

the incident photons, with frequency ν, expel the kth species of electron in the atom (possessing a binding or ionization energy I_k) with a kinetic energy of E_k.

The binding energy at any level for an atom is related to the atomic number and as the values for core electrons are fairly constant and characteristic for individual isotopes they can be used as the basis for the chemical analysis of surfaces as in X-ray **photoelectron spectroscopy**.

Binomial series. The series of values obtained by the expansion of the general expression $(x + y)^n$ for positive values of n.

$$(x + y)^n = x^n + nx^{n-1}y + \frac{n(n-1)}{2} x^{n-2}y^2 \ldots y^n$$

The general term ($m + 1$ term) for the expression is given by

$$\frac{n!}{m!(n-m)!} x^{n-m}y^m$$

The coefficients for the individual terms in the binomial expansion for increasing values of n correspond to the relative intensities of peaks in proton magnetic resonance multiplets arising as a result of first-order coupling. The values are best arranged in the form of the **Pascal triangle**.

Black-body radiation. Emission of radiation from incandescent solids depends upon the chemical composition and physical nature of the material. A black body is both a perfect absorber of radiant energy and a perfect radiator. Any radiation striking the black body is absorbed without loss due to reflection or transmission. The main interest in black-body radiation is in the distribution of the radiation emitted when the black body is heated, and this is taken as a standard for any source used to produce radiant energy for infrared spectrometers. Thus the suitability of the **Nernst glower** and the **globar** are assessed by comparison with the energy distribution from a black body as shown in the figure below.

The total emission of radiant energy is expressed by the **Stefan-**

Boltzmann law, while the **Wien laws**, the **Rayleigh and Jeans law** and the **Planck law** serve to describe other aspects of the relationship between energy, wavelength and temperature.

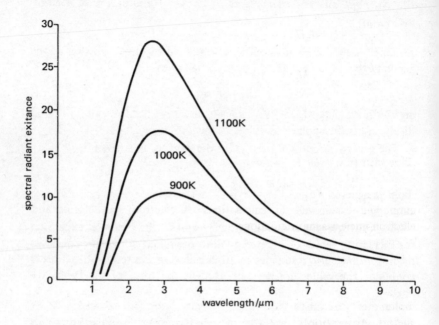

Figure 6. Distribution of black-body radiation

Blaze angle. In an echelette grating (*see* **Echelette and echelle gratings**) the faces of the grooves are cut at a constant angle to the plane of the original surface of the grating material. The angle of inclination of the grooves is known as the blaze angle.

Bloch equations. (n.m.r.). Development of the Bloch equations[41] assisted not only the calculation of magnetic moments in magnetic fields, but also the calculation of the shapes of absorption curves for simple n.m.r. spectra to be made.[42] When referred to a set of rotating axes the equations take the following form:

$$\frac{dM_x}{dt} = y(M_yH_0 + M_zH_1 \sin \omega_L t) - \frac{M_x}{T_2}$$

$$\frac{dM_y}{dt} = y(M_zH_1 \cos \omega_L t - M_xH_0) - \frac{M_y}{T_2}$$

$$\frac{dM_z}{dt} = y(-M_xH_1 \sin \omega_L t - M_yH_1 \cos \omega_L t) - \frac{M_z - H_0}{T_1}$$

in which M_x, M_y, M_z are magnet moments along axes x, y and z respectively, H_0 is the applied field and H_1 the oscillating field.

Blue shift (u.v.). *See* **Hyposochromic shift.**

Bohr magneton (μ_B) (e.s.r., n.m.r.). The unit of magnetic moment employed extensively in calculations of electron spin effects and electron-nuclear spin interactions.

$$\mu_B = \frac{eh}{4\pi m_e} = \frac{e\hbar}{2m_e} = 9.274\,078 \times 10^{-24} \text{ J T}^{-1}$$

Bolometer (thermistor) (i.r.). A detector used in infrared spectrometers. Absorption of radiation by the thermistor surface leads to a temperature change with a consequent alteration in the resistance. By incorporating the thermistor as one arm of a **Wheatstone bridge circuit** a current flow is obtained proportional to the amount of energy incident on the detector.[43]

The sensitive part of the thermistor is a 5 mm square flake, about 10 μm thick, made from a mixture of manganese, nickel and cobalt oxides.[44] The resistance change is about 4% for every degC rise in temperature. Highly sensitive bolometers suitable for operating at low temperatures have been made from superconductors such as columbian (niobium) nitride.[45]

Boltzmann constant (*k*). The ideal gas constant per molecule, the value being given by

$$k = \frac{R}{N_A} = 1.380\,662 \times 10^{-23} \text{ J K}^{-1}$$

$$= 1.380\,662 \times 10^{-16} \text{ g cm}^2 \text{ s}^{-2} \text{ deg}^{-1}$$

Boltzmann distribution. *See* **Maxwell–Bohzinann distribution and statistics.**

Boltzmann principle. This gives the statistical distribution of particles for any system subjected to thermal agitation under the influence of an electric, magnetic or gravitational field, such that the number of particles possessing a specified energy N_E is defined by the equation:

$$N_E = Ne^{-E/kT}$$

Boltzmann ratio. From the Boltzmann principle it can be shown that the ratio of the population in an upper energy state to that in a lower energy state is given by

$$\frac{N_E{}'}{N_E{}''} = e^{-h\nu/kT}$$

for nuclear energy levels this becomes

$$\frac{N_E{}'}{N_E{}''} = e^{-\mu H/IkT}$$

Bonding and anti-bonding orbitals (u.v.). The distribution and energy of electrons in atoms may be calculated from the **Schrödinger wave equation** and presented in terms of an orbital (a three-dimensional wave function) identified by the symbol ψ. Molecular orbitals may be considered to approximate to the sum or difference of the atomic orbitals of the constituent atoms, such that for a diatomic orbital the molecular orbital with the wave function ψ_{mo} can be of the two types:

$$\psi_{mo} = \psi_1 + \psi_2$$

and

$$\psi_{mo} = \psi_1 - \psi_2$$

where ψ_1 and ψ_2 are the wave functions for the atomic orbitals of the two atoms forming the bond.

If a molecular orbital is formed by the combination of two 1s atomic orbitals then its shape is ellipsoidal and it is uniformly distributed between the two nuclei producing a $1s\sigma$ orbital. This constitutes a bonding orbital and represents the lower energy state of the bond. In the second case the molecular orbital is concentrated outside the two nuclei and represents the higher energy anti-bonding $1s\sigma^*$ state. The types of orbitals formed by s and p atomic orbitals are shown in Figure 7, page 24. Similar, but more complex, orbitals are formed by d and f atomic orbitals.

Bouguer law (i.r., u.v.). *See* **Beer–Lambert law** and **Lambert law**.

Brackett spectral lines (em.). *See* **Atomic spectra**.

Bragg equation (pe.). When a beam of X-rays is directed into a crystal lattice it is diffracted from the atoms or ions forming the various planes in a manner that causes either interference or reinforcement of the beam. Reinforcement will occur when the equation:

$$n \lambda = 2d \sin \theta$$

is satisfied, where n is the order of the diffraction (n = 1, 2, 3 etc), d is the spacing between crystal planes and the θ the angle of the incident X-rays. This is, of course, similar to the equation for a diffraction grating (*see* **Gratings**). The peak reflectivity for X-rays can be of the order of 45 per cent and by using quartz crystals this diffraction procedure is

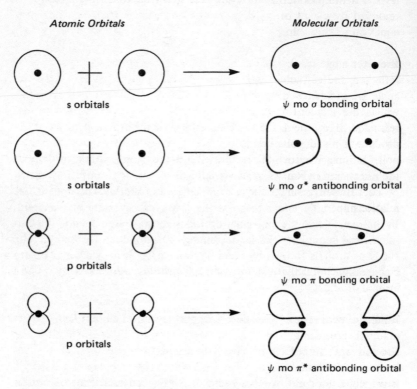

Figure 7. Bonding and antibonding orbitals

used as the basis for monochromation (*see* **Monochromator**) of hard X-rays for **photoelectron spectroscopy**.

Bremsstrahlung (pe.). The energy spectrum obtained from normal X-ray sources consists of a broad continuous band of radiation with characteristic emission lines superimposed upon it. The name 'Bremsstrahlung' has been given to this broad spectrum which has a maximum intensity of approximately two-thirds of the primary electron energy. As the presence of this radiation leads to an increase in the background level and the production of interfering peaks the undesired radiation is removed by means of an X-ray **monochromator** based upon the **Bragg equation** relationship.

Brewster angle (r.s.). Light incident upon a transparent surface is both reflected and refracted, the degree of each being dependent upon the angle of incidence of the light. The Brewster angle is the angle at which the incident light will be reflected if in one polarization and refracted if in the opposite polarization. This makes it possible to produce a plane polarized refracted ray from a light beam of mixed polarizations. Windows arranged at the Brewster angle are employed in the **lasers** used in **Raman spectroscopy**.

The Brewster angle for a substance can be calculated from the relationship:

$$\tan \theta_{\text{Brewster}} = n \text{ of the refracting material}$$

For example, for flint glass $n = 1.66$, hence the Brewster angle = $50°$ $56'$.

Brillouin zone (pe.). The complete energy system of a conducting material possessing 'free' electrons consists of a series of alternate allowed and forbidden bands. The permitted conduction bands are known as Brillouin zones. When Brillouin zones overlap the material is usually a good electron conductor and this is characteristic of metals. With materials that serve as insulators the zones are widely spaced by forbidden regions such that promotion of electrons from one Brillouin zone to another is only accomplished with great difficulty.

Broad-band n.m.r. Solids and viscous liquids give broad absorption peaks, when studied under n.m.r. conditions, that correspond to the total proton content of the material. Low-resolution broad-band n.m.r. spectroscopy has been developed for the determination of water and fat content in powders and seeds based upon this idea. The instruments for this purpose employ a large solid sample volume, 40 ml, in a 600–700 G (0.6–0.7 T) magnetic field with a r.f. of 2.7 MHz, and can determine liquid contents as low as 0.5% in the sample. The technique has been applied to the determination of H_2O in D_2O and to the liquid content in slurries[46] as well as for assessing the crystallinity of polymers.[47] In recent years improvements in instrumental techniques have made it possible to obtain n.m.r. spectra of solids with high levels of resolution. Particular interest has been focused on carbon-13 spectra[48] in which rapid sample spinning at 3kHz reduces the line broadening arising from chemical shift anisotropy.

Burner (a.s.). In many flame photometers the steady flame needed for the atomic spectra is obtained by using a modified Meker burner. The **fuel gas** is passed through a small orifice and mixed with a large amount of air before being burnt above the metal grill at the top of the burner. After nebulization the sample solution is introduced directly into the flame.

For **atomic absorption spectroscopy** the burner is a long horizontal tube with a narrow slit along its length. This produces a thin flame with a long path length which can be turned into or away from the source of radiant energy. *See also* **Nebulizer**.

C

Calibration lines (em.). *See* **R.U.**

Calibration peaks (Marker peaks) (i.r., u.v.). To correct for any errors in spectral chart alignment and for spectral deviations due to instrumental variations it is common practice to use a standard compound to place known absorption peaks as markers on the spectra of materials under examination.

An extensive list of spectral standards has been compiled for the calibration of spectra in this way;[49,50] the list includes the characteristic absorptions for elements such as neon and argon as well as simple molecules like carbon dioxide, acetylene and nitric oxide.

The most common substance employed as a marker for infrared spectra is polystyrene; strong peaks at 3027 cm^{-1} (3.30 μm), 2851 cm^{-1} (3.51 μm), 1602 cm^{-1} (6.24 μm), 1028 cm^{-1} (9.73 μm) and 907 cm^{-1} (11.03 μm) being of particular use for this purpose. Ammonia and indene have also been used.

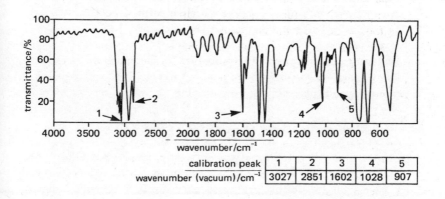

calibration peak	1	2	3	4	5
wavenumber (vacuum)/cm^{-1}	3027	2851	1602	1028	907

Figure 8. Calibration peak spectrum of polystyrene

27

Carbon filament atom reservoir (a.s.). The development of carbon filament and carbon tube atom reservoirs was the first step in the growth
of flameless techniques for the atomization of samples for atomic
absorption and atomic fluorescence spectrometry. As a result **electrothermal atomizers** are commonly employed now for this purpose as
they enable materials to be examined without the use of highly flam-
mable solvents and obviate the use of **nebulizer** sampling systems.

In a typical system, shown in Figure 9, the sample (2–50 μL) is
deposited in the centre of a graphite filament and evaporated by passing
a current through the graphite. It is then atomized, in an inert atmosphere at 2500 °C by passing a large current (20–500 A) through
the graphite. Signals produced by this method can last for several
seconds.[51]

Graphite rods and tubes may be used hundreds of times. By their
use flameless atomic absorption has increased sensitivities by several
orders of magnitude and is frequently superior to neutron activation
analysis.[52] *See also* **Delves cup**; **Massman furnace**; **Tantalum boat**.

Cartesian coordinates. Points or objects in three dimensions can be
located by reference to the scales of three orthogonal axes, the Cartesian coordinates. These mutually perpendicular axes are usually de-
noted as x, y and z such that the location of a point has its position
defined by a value of x, y and z measured with respect to each axis.
They are named after the mathematician Descartes.

Cascade process (a.s., u.v.). The procedure for increasing the strength of
a weak signal by the progressive build-up of electron displacement in a
series of **dynode** plates in a **photomultiplier**.

C.A.T. (n.m.r.). *See* **Computer of average transients**.

Cathode (m.s., pe.). The negative electrode in an electron tube or **electron bombardment ion source**, which acts as an emitter of electrons and
as a collector for positively charged particles. In a mass spectrometer
it is the source of electrons for the ionization and fragmentation
processes. In this case the **accelerating slits** must be negative with

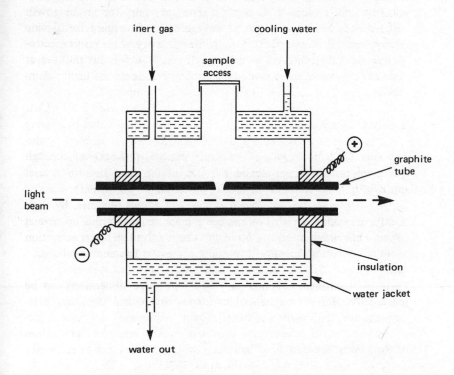

Figure 9. Graphite rod electrothermal
atomization

respect to the cathode in order to attract the positive ions into the
analyser tube. For the generation of X-rays the cathodes are required
to produce high speed electrons and for this purpose tungsten, thoria
and rhenium are commonly used as they are good thermionic materials
and can withstand high operating temperatures. *See also* **Electron
gun.**

Cauchy distribution. *See* **Lorentzian curve.**

Cavity cell (e.s.r.). The absorption cell most commonly employed in electron spin resonance studies is a resonant cavity formed by closing off the ends of a short length of **waveguide** cut to a simple multiple of the wavelength dimension of the guide. This leads to the microwave power being concentrated in the cavity by reflection between the ends. The sample is contained in a quartz tube inserted into the centre of the cavity.

Cavity cell (i.r.). *See* **Microcells**.

C bands (u.v.). Moser and Kohlenberg[53] designated the absorption bands occurring on the longer-wavelength side of the main absorption band for conjugated systems as the C bands. *See* **Bands (ultraviolet)**.

Centre of symmetry (i.r.). A molecule possesses a centre of symmetry when a line drawn from any atom and passing through the centre of the molecule meets an identical atom in a corresponding position diametrically opposite. *See also* **Axis of symmetry** and **Plane of symmetry**.

c.g.s. units. A system of units based upon the centimetre, gramme and second now obsolescent and superseded by **SI units**.

Charge on an electron. *See* **Electron**.

Charge on a proton. *See* **Proton**.

Chemical actinometer (pl.). Rates of photochemical reactions are studied by the use of this special type of detector system. Essentially it is a cell containing a chemical compound which undergoes a reaction on exposure to light. The extent of the reaction is a measure of the quantity of radiation to which it has been subjected. One substance used for this purpose, covering the wavelength range 250–500 nm is potassium trisoxalato-ferrate(III) trihydrate, $K_3 Fe(C_2 O_4)_3 . 3H_2 O$, in sulphuric acid.[54] The photochemical reaction produces iron(II) ions which are measured quantitatively as the 1,10-phenanthroline complex. This can detect 5×10^{-10} N quanta and is accurate to ± 2%.

Chemical actinometers have also been developed in the form of transparent films[55] incorporating o-nitrobenzaldehyde. This system is useful for studying radiation intensities between 300 and 410 nm.

Chemical ionization (CI) (m.s.). In chemical ionization the sample is mixed with excess of a reagent gas (normally methane) before being introduced into the mass spectrometer ionization chamber.[56] The methane is preferentially ionized whilst the direct ionization effect on the sample is negligible. However, the molecular and combination ions from the methane 'pass on' the ionization to the sample molecules. As a result fragmentation is more gentle and selective compared with **electron** impact **ionization**. Strong M (molecular) or $M + 1$ (following hydride extraction) peaks are obtained by this procedure. *See also* **Field desorption** and **Field ionization**.

Chemical shift (n.m.r.). The displacement between the absorption peak of the nucleus of interest and the absorption peak of an appropriate reference standard in an n.m.r. spectrum. In proton studies the sharp absorption band of the methyl groups of tetramethylsilane (*see* **TMS**), included as an internal standard, is the reference peak against which the chemical shifts of other peaks are measured. For phosphorus nuclei the reference is normally phosphoric acid.

In quantitative terms the difference may be measured in **hertz,** or in δ or τ **parts per million**.[57] As the chemical shift is proportional to the field strength, comparison of values obtained on different instruments is best achieved by using the dimensionless units of parts per million. τ values are specifically referred to TMS in proton spectra.

Chemical shift values obtained for different protons are dependent upon the individual chemical and magnetic environments; in general a high electron density around the protons leads to a small chemical shift and vice versa. The following table gives an indication of the range of the main chemical shift values. *See also* **Deshielding** and **Shielding**.

Chemical shift (pe.). Just as with nuclear magnetic resonance, the chemical environment of an atom has a small but significant influence upon the actual value obtained for the electron **binding energy**. This

Table 3

Proton Group	Chemical Shift Range
	(ppm δ)
CH_3-C	0–2
CH_3-C- $\overset{\parallel}{O}$	2–3
$HC{\equiv}C-$	2.5–3
$-CH_2-$	2.8–5.5
$\overset{H}{\underset{H}{}}C=C\overset{}{\underset{}{}}$	4.5–7
(benzene ring)	6.5–8.5
$-C\overset{H}{\underset{O}{}}$	9–10
$-C\overset{O}{\underset{OH}{}}$	10–13
$-OH$	1–8

means that two identical atoms in the same molecule but forming distinctly different parts of the structure will have binding energies differing by possibly 3–4 eV. This variation is also known as the 'chemical shift' and the range of values of binding energies for many elements in different compounds is now available in tables.[58] These chemical shift values have been used as a basis for qualitative analysis of compounds but the identifications are subject to some uncertainties due to the relatively large errors (± 0.2 eV) in the experimental values obtained for the binding energies.

Chemical shift reagents (n.m.r.). Since the original observation of the effect on cholesterol[59] a number of paramagnetic lanthanide complexes have been found in which the central lanthanide ion can increase its coordination number by interaction with **lone pair electrons** in functional groups of organic compounds. The result of this is that the n.m.r. spectrum of organic compound can be altered and significant changes in **chemical shifts** may be observed. The method can be used to obtain greater information about the molecular structure, especially in cases in which the normal n.m.r. spectrum possessing overlapping absorption signals is separated by the action of the shift reagent into clearly defined multiplets.[60,61]

First-order spectra have been obtained by the use of tris-(dipivalomethanato) europium, $Eu(DPM)_3$, whilst other work[62] has shown that tris-$(2, 2, 6, 6$-tetramethylheptane-$3, 5$-dionato)-praseodymium, $Pr(tmhd)_3$, is of similar value for both ^1H and ^{13}C n.m.r. spectra.

Chemiluminescence (pl.). **Luminescence** occurring during the course of chemical reactions[63] which arises from the emission of radiation from molecules that have absorbed energy during the course of the chemical reaction.

Chopper. *See* **Rotating sector mirror.**

Chromatic aberration. As the refractive index for any transparent substance differs for the different wavelengths of the light the individual wavelengths are dispersed by varying amounts. As a result white light can be dispersed into its individual wavelengths including the infrared, ultraviolet and visible spectra (*see* **Infrared spectra** and **Ultraviolet/visible spectrum**). Images produced by lenses will, therefore, possess coloured edges due to the chromatic aberration of the lens. *See also* **Dispersion.**

Chromophore (u.v.). Chemical functional groups which are capable of exhibiting characteristic absorptions in the ultraviolet and visible regions of the spectrum even when they are not in conjugation with or affected by the influence of other groups. Most chromophores possess some degree of unsaturation, the carbonyl groups being typical and exhibiting absorptions between 160 and 170 nm and between 180 and 190 nm. These absorptions occur due to π to π^* and n to π^* transitions, typical transitions being listed in Table 4. Chromophoric absorption values are greatly enhanced when conjugation exists between chromophores or when they are influenced by **auxochromes**. *See also* **Bonding and anti-bonding orbitals** and **Hyperchromic effect**.

Table 4. Typical Chromophores.

Group	Transition	Approximate wavelength (nm)
>C=C<	$\pi \rightarrow \pi^*$	175
$-C\equiv C-$	$\pi \rightarrow \pi^*$	190
>C=O	$\pi \rightarrow \pi^*$	180
>C=O	$n \rightarrow \pi^*$	280
$-C\equiv N$	$\pi \rightarrow \pi^*$	190
$-C\equiv N$	$n \rightarrow \pi^*$	300
>C=S	$n \rightarrow \pi^*$	500
$-NO_2$	$\pi \rightarrow \pi^*$	210
$-NO_2$	$n \rightarrow \pi^*$	275

Circular dichroism. Optically active materials frequently exhibit different **molar absorption coefficients** for left and right circularly polarized light. This is known as circular dichroism and varies with the change in wavelength. The difference between the two values for ϵ over a range of wavelengths gives a plot of $\epsilon_d - \epsilon_1$ which is bell shaped and may have either positive or negative values.[64] The phenomenon has been used as the basis for both qualitative and quantative analysis, for example, for alkaloids in buffered media.[65]

Collimator. This part of the optical system of a photometer is a lens or mirror arranged to produce a parallel beam of light to or from the prism or diffraction grating. For this purpose the radiation source is placed at the focal point of a divergent lens; alternatively this can be achieved by passing the radiation through a narrow slit at the focal point of the lens.

Combination lines (n.m.r.). Combination lines occur when two or more nuclei undergo simultaneous changes of spin states (*See* **First-order spin pattern**), giving rise to overlapping peaks in the n.m.r. spectrum. First-order spectra do not usually have peaks due to combination transitions although these can appear under saturation conditions.

Combination modes or vibrations (i.r.). A number of weak absorptions occur in infrared spectra corresponding to the sum of two or more fundamental vibrational frequencies. These combination modes arise from the anharmonicities of the oscillators which lead to an interaction of the vibrational states in polyatomic molecules. The resulting absorptions are weak compared with those due to fundamental vibrations and **overtones**. *See also* **Difference modes**.

Computer of average transients (C.A.T.) (n.m.r.). A device originally employed in n.m.r. spectroscopy but now used on a wider basis to improve **signal to noise ratios** by increasing the size of weak signals compared to the general noise level. This is done by a scan averaging procedure carried out as spectra are repeatedly scanned and stored in the memory of a small computer. For 'n' scans the increase of the signal size over the noise is $\sqrt{n}$. This is because for n scans the signal increases n times while the noise, which is random, increases $\sqrt{n}$ times and the nett improvement is $n/\sqrt{n}$ or $\sqrt{n}$. Thus a spectrum scanned 49 times gives a seven-fold improvement over a single scan.

Contact hyperfine interaction (e.s.r.). *See* **Fermi contact interaction**.

Continuous laser (r.s.). *See* **Laser**.

Continuous mode (m.s.). A method for the continuous production of positive ions in the **time-of-flight mass spectrometer**. Ionization and fragmentation of molecules is carried out by a continuous stream of electrons throughout the cycle of operations in order to produce a constant supply of positive ions. The ions are only attracted away during the drawout pulse of the instrument. *See also* **Pulsed mode**.

Continuous spectra (em.). Incandescent solids emit continuous spectra over a broad band usually with the absence of any sharply defined lines. Materials producing such spectra include ceramics and fused silicacious mixtures. In some cases the broad emission of radiation approximates to that from a black body (*see* **Black-body radiation**) and such substances can be used as radiation sources for infrared spectroscopy. *See also* **Globar**; **Nernst glower**; **Opperman source**.

Contour (n.m.r.). *See* **Curvature**.

Core electron (pe.). The major application of X-ray **photoelectron spectroscopy** has been in the study of the electrons occupying the inner orbitals of the atomic structures and that are not normally involved in the formation of chemical bonds. The investigations on these 'core electrons' form the basis of electron spectroscopy for chemical analysis (ESCA).

Core hole (pe.). When a **core electron** is removed, by any means, from an atom it temporarily leaves a vacancy known as a core hole. This core hole is normally filled fairly rapidly by an electron falling from a higher energy level, in the process energy is released by one of three methods: (i) by **X-ray fluorescence**; (ii) by the release of a secondary electron by the **Auger effect**; (iii) by a **Coster–Kronig process**.

Cornu mounting. The prism system employed in those spectrophotometers in which radiation is passed in through one face of the prism and out through the opposite face, as shown in Figure 10. The prism employed for this purpose is formed from two 30° quartz prisms, one of right-handed quartz and the other of left-handed quartz, joined to give a 60° prism producing admirable dispersion but no polarization.

Coster–Kronig process (pe.). A process similar to the **Auger effect** in which a vacancy in an inner electron orbital is filled by transition of an electron from another outer orbital and the release of an Auger electron. In the Coster–Kronig process the same type of double vacancy is created, but in this instance one of the vacancies is in a shell with the

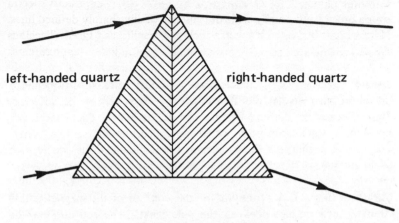

left-handed quartz   right-handed quartz

Figure 10. Cornu prism

same principal **quantum number** as that from which the initial ionization occurred. Electrons emitted by these processes possess only low kinetic energies and are not greatly studied in **photoelectron spectroscopy**.

Coupling constant (a_i) (e.s.r.). Hyperfine splitting in electron spin resonance spectra occurs as a result of the interaction of atomic nuclei with the electrons of the system under examination. The coupling constant is the distance between the individual peaks in the hyperfine splitting multiplet; its values is determined by the extent of coupling arising from the magnetic moments of the interacting nuclei and is expressed in gauss.

Coupling constant (J) (n.m.r.). The coupling constant (measured in hertz) is the separation between the peaks of a first-order multiplet produced as a result of **spin–spin coupling**. The larger the value of J, the greater is the coupling between the nuclei. The value of J is independent of the operating frequency of the instrument upon which it is measured.

Cracking pattern (m.s.). A name for the characteristic fragmentation spectrum obtained for a substance by mass spectrometry. *See* **Mass spectrum**.

Crossed-coil method (n.m.r.). *See* **Induction method**.

Crystal detector (e.s.r.). In electron spin resonance spectrometers a silicon-tungsten crystal rectifier is usually employed as the detector. This is located at the end of one arm of the **hybrid T** and receives a signal only when absorption of energy occurs in the sample cavity. The crystal produces a direct current proportional to the square root of the microwave signal.

Curvature (n.m.r.). A synonym for the contour of the magnetic field strength of a magnet between the pole pieces. The contour may be spoken of as 'dome-shaped', 'dish-shaped' or 'flat'. Absorption peaks will be distorted unless the contour is flat. The curvature of the field is adjusted during the tuning up procedure and is carried out roughly by **cycling** the electromagnet.

Cuvette. A name given to the glass or silica sample tubes used for the study of liquids and solutions in ultraviolet and fluorescent spectroscopy.

Cycles per second (n.m.r.). *See* **Hertz**.

Cycling (n.m.r.). A rough adjustment of the **curvature** of the magnetic field strength between the pole pieces of an electromagnet is obtained by passing an abnormally large current through the coils of the electromagnet for two or three minutes followed by a reduction to the normal operating current level for the instrument. The operation is intended to produce a flatter contour of the magnetic field strength between the poles. *See also* **Shim coils**.

Czerny–Turner mounting. By using two concave mirrors with a plane diffraction grating a **stigmatic** and achromatic optical system (*see* **Achromatic lens**) is produced. By this means all wavelengths are brought to a focus without changing the detector to mirror distance in the spectrometer. Any change in wavelength is simply obtained by rotating the grating. This type of system[66,67] is a modification of that described by Ebert[68] in 1889. The Czerny-Turner mounting is now used for monochromation in a wide range of spectrometers including those for emission and Raman spectroscopy.

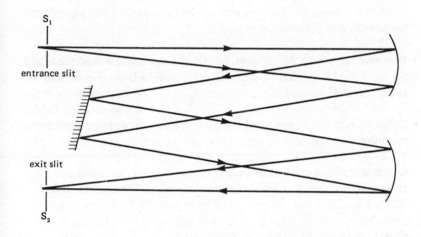

Figure 11. Czerny-Turner mounting

D

Dark current. The background current that flows in photoemissive and photoconductive detectors, such as those used in ultraviolet and infrared spectrophotometers, when no radiant energy is falling upon the detector. It is a temperature-dependent effect and in manual single-beam instruments must be corrected for before taking a reading on every occasion. Operation of these detectors at low temperatures ($- 20\,°C$) reduces the dark current by many orders of magnitude compared with that at room temperature.

Daughter ion (m.s.). In a transition involving **metastable ions** the actual ion recorded by the detector, although its mass/charge value is recorded as the value M^* instead of the correct M_B.

D bands (u.v.). Within the ultraviolet region the weakest absorption band for conjugated compounds occurs at the longest wavelengths. Moser and Kohlenberg[69] designated these as the D bands. *See* **Bands (ultraviolet).**

d.c. arc source (em.). Samples can be vaporized for emission spectroscopy by subjecting them to a d.c. arc produced by a voltage of up to 300 V, and leads to electronic transitions in elements. The temperature within the arc may be as high as 8000 K. In most instruments the arc is struck between graphite electrodes. The d.c. arc is particularly useful for the study of neutral atoms present in small amounts. *See also* **a.c. arc source** and **a.c. spark source.**

de Broglie relationship. All matter is believed to possess both a particle property and a wave property. The two are related according to the de Broglie relationship in which the momentum of the particle is equal to the **Plack constant** divided by the wavelength:

$$mv = \frac{h}{\lambda} = h\tilde{\nu}$$

40

Decoupling (n.m.r.). *See* **Multiple resonance.**

Deformation vibrations (i.r.). *See* **Bending vibrations.**

Degeneracy. A term used when one energy level corresponds to two or more states of motion. It arises when the symmetry of a molecule is such that certain fundamental frequencies are equal and is a common feature in infrared spectroscopy. When two such frequencies are equal the system is termed double degenerate and when three frequencies are equal it is triply degenerate.

Degrees of freedom (i.r., r.s.). For a molecule of n atoms there exist a total of $3n$ degrees of freedom. In the case of a non-linear molecule there are three translational degrees of freedom and three rotational degrees of freedom leaving a total of $3n-6$ vibrational degrees of freedom. For a linear molecule the number of vibrational degrees of freedom is $3n-5$. Vibrational degrees of freedom may be further subdivided into **bending vibrations** and **stretching vibrations.**

Delta values (n.m.r.). *See* **Parts per million** and **Chemical shift.**

Delves cup (a.s.). Very rapid determinations of the lead content of blood samples are obtained by introducing the sample contained in a small nickel cup[70],[71] directly into the non-luminous air–acetylene flame of an atomic absorption spectrometer. The blood samples are partially oxidized by hydrogen peroxide before the analysis which takes about five minutes from beginning to end, the actual atomic absorption reading only requiring a few seconds. The procedure has also been used for the analysis of paint and milk.[72]

Densitometer comparator (em.). In both qualitative and quantitative emission spectroscopy the nature of unknown substances is determined by comparison with a series of standard spectra. This is done using special comparators in which two plates can be studied side by side. In the most modern forms of instrument the two plates are illuminated, and enlarged images projected onto a screen as illustrated in Figure 12.

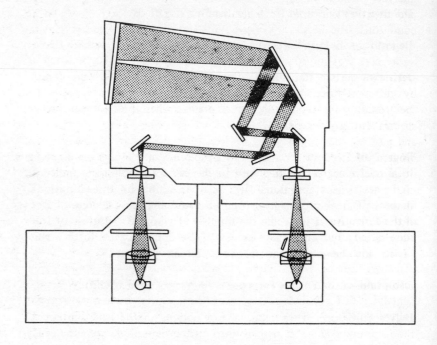

Figure 12. Optical arrangement of a projection
comparator

Depolarization factor (or ratio) (ρ_N) (r.s.). By enclosing the sample
tube in polarization sheets each line in a Raman spectrum can be split
into a horizontal component (I_h) and a vertical component (I_v). The
depolarization factor (ρ_N) is the ratio of these two components. As
I_v is always weaker than I_h the value of ρ_N is less than unity. The
theoretical limit for the value of ρ_N is 0.857.

$$\rho_N = \frac{I_v}{I_h}$$

Measurement of the depolarization factor is carried out by placing an analyzer prism in the Raman beam and setting first horizontally (at 0°) and then vertically (at 90°) while measuring the spectrum under both conditions. The depolarization factor is then obtained by calculating the ratios of the band intensities from the two spectra.

Derivative spectra. The derivative curve for any distribution is obtained by differentiating the equation for the particular distribution. In spectroscopy the technique is used to magnify the fine structure of spectra and involves calculating first, second or higher order derivatives and plotting these in addition to or instead of the normal spectrum.[73] Modern electronics have made it possible to obtain automatic plotting of derivative curves and it is this that has led to the growth of interest in the technique. The main advantage of the derivative approach lies in the enhanced spectral features that are obtained, and as the order of the derivative increases so this advantage grows although the **signal to noise ratio** is found to decrease.

Electron spin resonance spectra are normally recorded as first-derivative curves obtained by applying a modulating field to the absorption current. The shapes of the various derivative curves obtained by differentiating a Gaussian (normal) distribution are shown in Figure 13.

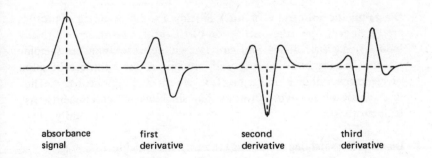

| absorbance signal | first derivative | second derivative | third derivative |

Figure 13. Derivative plots

Deshielding (n.m.r.). Absorption peaks on n.m.r. spectra that occur at low field strength (away from the TMS signal) are said to be deshielded if they absorb at low field on the n.m.r. spectrum as a result of being in a position in which the induced magnetic field produced by the electron circulation acts in the same direction as the applied field. Protons in the —CHO group of aldehydes are typical of deshielded atoms as a result of the electron circulation of the electrons forming the carbonyl double bond. *See also* **Chemical shift**; **Ring current effect**; **Shielding**.

Detectors. *See* **Barrier-layer cell**; **Bolometer**; **Photoelectric infrared detectors**; **Photoemissive tube**; **Photomultiplier**; **Pneumatic cell**; **Thermocouple**.

Deuterium discharge lamp (u.v.). *See* **Hydrogen discharge lamp**.

Deuterium exchange (n.m.r.). The presence of exchangeable protons, such as those attached to oxygen and nitrogen, can be shown in n.m.r. spectra by their disappearance from the spectrum after the sample has been shaken with D_2O. This has the advantage that in cases in which the absorptions of the exchangeable protons overlap with those of other protons the spectrum may be substantially simplified by deuterium exchange. The resulting peak for the HDO molecules formed (arising near to 5δ) can be used to estimate the number of exchangeable protons in the molecule.

Diagrammatic anisotropy (n.m.r.). **Shielding** and **deshielding** of protons are dependent upon the orientation of the protons with respect to any induced magnetic fields and are known collectively as anisotropic effects. Diamagnetic anisotropy is brought about by the circulation of electrons within π and σ electron clouds and leads to deshielding in the case of aldehydic protons, but shielding in the case of acetylenic protons.

Diamagnetic shielding (screening) (n.m.r.). *See* **Shielding**.

Diamagnetism. The magnetic property possessed by a material as a result of the circular of electrons. The magnetism produced acts in opposition to that of the applied field. Where unpaired electrons exist in a substance **paramagnetism** arises and this is dominant over any diamagnetism.

Difference modes (i.r.). Some of the absorptions recorded in infrared spectra correspond to the difference between two vibrational frequencies. In these cases the molecule already existing in one excited vibrational state absorbs enough additional radiant energy to raise it to another vibrational level in a different vibrational mode. The measured absorption is then the difference between the two. *See also* **Combination modes**.

Diffraction. Diffraction occurs when a wavefront passes through narrow slits and holes and at sharp edges. In the case of a number of parallel slits the effect is to produce a series of wavefronts which are refracted away from the original line of direction. This idea is the basis of the diffraction grating (*see* **Gratings**) used extensively in spectrophotometers, although most instruments now use the echelette reflectance gratings.

Diffraction gratings. *See* **Gratings**.

α, α-**diphenyl-β-picrylhydrazyl** (e.s.r.). *See* **DPPH**.

Dipole–dipole broadening (n.m.r.). A broadening of n.m.r. absorption peaks occurs both in the solid state and in viscous liquids as a result of the magnetic dipoles because the magnetic field strength varies throughout such media. This variation arises because the physical nature of the material prevents the averaging out of the magnetic fields due to the nuclei. *See also* **Broad-band n.m.r.**

Direct inlet system (m.s.). Substances possessing low vapour pressures can be introduced into a mass spectrometer, in the absence of a probe inlet device, by bleeding directly from the sample holder into the ionization chamber without using an intermediate reservoir expansion chamber or leak valve. The system is unsuitable for any substance possessing a high volatility. *See also* **Batch inlet sampling system** and **Direct insertion probe.**

Direct insertion probe (m.s.). Solids and involatile liquid samples can be vaporized into the ionization chamber of the mass spectrometer by placing them in a small glass sample tube on the end of a rod that can be inserted into the instrument, through a vacuum lock, to a position close to the ionizing beam. Direct insertion probes can be heated up to 350 °C by means of a filament around the sample holder, and some can also be operated at temperatures as low as − 50 °C by passing liquid N_2 through the probe. *See also* **Batch inlet sampling system** and **Direct inlet system.**

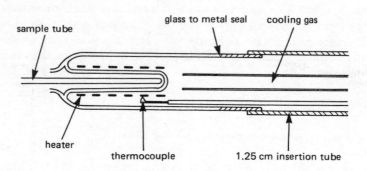

Figure 14. Direct insertion probe.

Dispersion. In its loosest sense this expression is employed to describe the ability of any transparent substance to separate the wavelengths of radiant energy passing through, as illustrated in Figure 15. In spectroscopy it is used with special reference to the prisms (*see* **Prisms, infrared** and **Prisms, ultraviolet**) and **gratings** employed for the monochromation of ultraviolet, visible and infrared radiation.

For very small slit widths dispersion depends only on the refractive index of the material and the size of the prism. The dispersive power for a prism is given by $\partial n/\partial \lambda$ and is used in the calculation of the resolution (*see* **Resolution: spectrophotometers**).

For gratings the dispersion is given by $\partial \theta/\partial \lambda = n/a \cos \theta$ where a is the distance between adjacent lines, θ the angle of reflectance and n the order of the spectrum. If θ is small $\cos \theta \approx 1$ and $\partial \theta/\partial \lambda = n/a$.

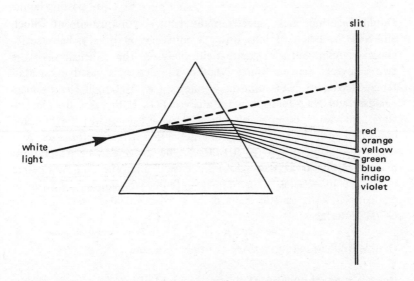

Figure 15. Dispersion by a prism

Doppler broadening (a.s., em.). One of the causes of the broadening of absorption and emission spectral lines is the thermal motion of the atoms. Doppler broadening is the name given to this effect and is proportional to $(T/A_r)^{1/2}$. *See also* **Pressure broadening**.

Doppler effect. The apparent change in frequency of a wave motion passing through a stationary medium, as determined by the detector, as a result of the relative movement of the source towards or away from the detector. In spectroscopy any moving source will give rise to a Doppler shift and this is one of the features of **Mössbauer spectroscopy.** The shift has also been detected in the spectra of moving stars.

Line broadening in the spectra of gases occurs because of the different speeds of the individual gas molecules giving rise to the Doppler effect.

Double-coil method (n.m.r.). *See* **Induction method.**

Double-focusing mass spectrometer (m.s.). An instrument which analyses the beam of ions using a combination of a radial electrostatic analyser and a magnetic field analyser. The dimensions of the two analysers depends upon whether the design is based upon Mattauch–Herzog or Nier–Johnson geometry (*see* **Mattauch–Herzog mass analyser** and **Nier–Johnson mass analyser**). In both cases the electrostatic analyser acts as an energy selector and improves the focusing compared with the simple **magnetic deflection mass spectrometer.** As a result resolution of up to 20 000 can be obtained (*see* **Resolution: mass spectrometry**). Many double-focusing instruments now operate using two ion beams[74] that enable reference materials to be run in comparison with samples and eluates from GC/MS interfacing (*see* **GC/MS interfaces**).

Double irradiation (n.m.r.). *See* **Multiple resonance.**

Double monochromator (i.r., r.s., u.v.). The use of double monochromation has become a standard feature in many spectrophotometers. Originally it involved the use of a prism to remove the unwanted higher orders of diffraction produced by **gratings.** In modern instruments the double monochromation is achieved by two coupled and synchronized diffraction gratings arranged in a double Czerny–Turner configuration (*see* **Czerny–Turner mounting**) such that the light from the first monochromator is reflected and passed through a slit which serves as the entrance slit for the second monochromator.

Double monochromation gives improved resolution and a reduction in stray radiation, but there is some loss of energy due to increased path length and absorption.

Double resonance (n.m.r.). *See* **INDOR** and **Multiple resonance**.

Doubly charged ions (m.s.). Some of the positively charged ionic species detected in mass spectrometry are found to carry a double charge (and sometimes even more). These give rise to mass/charge peaks at half the true mass value for the ions, as in these cases the instrument is detecting $m/2e$. For ions possessing an odd true mass value the peak for the doubly charged ion is found as having an apparent half mass unit falling between two integral mass/charge values.

DPPH (e.s.r.). This is the common standard used for comparison in electron spin resonance studies as it exists completely in the free radical state. It contains 1.63×10^{21} unpaired electrons per gram and can be diluted to give an e.s.r. signal corresponding to any required number of electrons by using carbon black as a diluent. It can also be used as a solution in benzene.

α, α-diphenyl-β-picryhydrazyl

Drift region (m.s.). In **time-of-flight mass spectrometers** the charged ions pass along a straight tube about two metres long. This is the drift region of the instrument and the period of time spent by the ions in the tube increases with the mass of the ion.

DSS (n.m.r.). Sodium 2, 2 - dimethyl - 2 - silapentane - 5 - sulphonate $(CH_3)_3 SiCH_2 CH_2 SO_3 Na$ is an alternative reference to TMS for proton n.m.r. Its main advantage is that it is a water-soluble solid and can be used as an **internal standard** in aqueous and D_2O solutions, with which TMS is immiscible.

Dynamic nuclear polarization (n.m.r.). *See* **Overhauser effect.**

Dynode. An electron-sensitive plate used in **photomultipliers.** Amplification of signals is achieved by directing photoelectrons onto the dynode plate which is designed to give a multiplication factor of at least two for every primary electron falling on it. The plate is most commonly made from a copper-beryllium or silver-magnesium alloy. A series of dynodes is employed in photomultiplier tubes to produce a **cascade process.**

E

E bands (u.v.). Bowden and Braude[75] assigned the letter E to the weak ultraviolet absorptions arising from the movement of electrons in aromatic systems. These are only measured in the vacuum u.v. region below 200 nm. This system of nomenclature conflicts with that used by other authors. *See* **Bands (ultraviolet)**.

Echelette and echelle gratings. The echelette grating is a very accurately ruled reflection **grating** with about 1000 lines cm^{-1} and a broad groove such that about 75 per cent of the reflected light is concentrated in one order.

An echelle grating is a coarse form of echelette grating with fewer fine lines which are more widely spaced. This arrangement gives a higher resolution over a narrow waveband such that 50 lines cm^{-1} can give a resolving power comparable with 10 000 lines cm^{-1} on a transmission diffraction grating.

Effective bandwidth (i.r., u.v.). Although the ideal **monochromator** should produce a line of single wavelength this perfect state is never realised in practice. Instead of the single isolated value there is always a finite width to the transmitted wavelength band spread on either side of the required value, and the 'effective bandwidth' is the width of this transmitted band measured at half the maximum transmittance value. This serves as a basis for comparison between monochromators as the narrower the effective bandwidth, the greater is the degree of monochromation. For work in the ultraviolet region the effective bandwidth needs to be < 1 nm. Figure 16, page 52, shows the inter-relationship between the effective bandwidth, the required monochromatic wavelength and the maximum per cent transmittance for the monochromator.

Eigenfunction (Ψ). This is a position-dependent amplitude and when multiplied by its complex conjugate Ψ^* it gives the probability

51

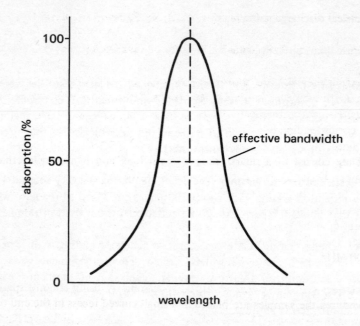

Figure 16. Effective bandwidth

$$P = \Psi * \Psi$$

that a given particle occupies a particular defined position. An eigen-function must be an acceptable solution of the wave equation. *See* **Schrödinger wave equation.**

Eigenvalues. Values obtained from solving the **Schrödinger wave equation** for eigenfunctions of a system.

Einstein (E). The amount of energy absorbed by one mole of material undergoing a photochemical reaction based upon the **Stark–Einstein law.**

$$E = N_A h\nu$$

Electrical discharge getter pumps (m.s.). *See* **Sputter-ion pump**.

Electric quadrupole broadening (n.m.r.). *See* **Quadrupole broadening**.

Electrodeless discharge tubes (a.s.). Sources for atomic absorption and atomic fluorescence originally developed for alkali metal determinations. Radiation intensities obtained are much greater than those from the corresponding hollow **cathode** lamps and produce very sharp spectral lines.[76]

They consist of a quartz tube 2-7 cm long and 8 mm in internal diameter containing up to 20 mg of the require element (identical with that to be studied) and sometimes an equal amount of iodine. The tubes are filled with argon at low pressure (about 2 torr).[77] Under the operating conditions all the charge in the tube should be in the vapour state. The tubes are subjected to a microwave frequency of 2000 to 3000 MHz.

Electrodes (em.). For most emission spectrometry using arc and spark procedures the samples are placed in a small curved recess in the end of a graphite electrode. The second, upper, electrode is a similar graphite rod shaped to a point and aligned with the sample.

In some cases, where the sample is a solid conductor, the electrodes may be made from the sample itself. Metallic electrodes have been used for general determinations but are best employed for providing reference spectra of themselves.

A special type of electrode that can be reused one hundred times has been developed[78] from vitreous carbon, and is called the glassy electrode. It is claimed to give emission spectra three times as intense as those obtained from graphite electrodes and is less prone to contamination owing to its pore-free surface.[79] The glassy electrode enables metals in solution to be determined with a greater precision.

Electromagnetic spectrum. For most spectroscopic purposes the electromagnetic spectrum is considered to consist of the region of radiant energy ranging from wavelengths of 10 metres to 1×10^{-10} centimetres. Only certain portions of this range are currently used for spectroscopic measurements, although many wavelengths are used for such purposes as broadcasting and radar. The diagram below illustrates the scope of the present applications.

Figure 17 here

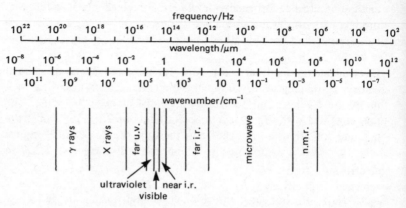

Figure 17. The electromagnetic spctrum

Electron. This name is most commonly used for the negatively charged elementary particle more specifically called the negatron. The mass (at rest) is $m_e = 9.109\,534 \times 10^{-31}$ kg with a charge of $1.602\,1892 \times 10^{-19}$ C, and a spin quantum number of ½.

The positive electron, possessing identical values but an opposite charge, is called the positron.

Electron bombardment ion source (m.s.). The formation of positive ions by electron bombardment of molecules is the most common form of ion source used in mass spectrometry and is known as electron ionization (EI). The molecules are passed into the ionization chamber of the spectrometer which is at a temperature of about 200 °C and a pressure of about 10^{-7} torr, where they are bombarded at right angles to their line of direction by a beam of low-energy electrons. The electrons for this bombardment are produced from an incandescent tungsten or rhenium filament heated to 2000 °C with energies covering the range 10–100 eV.

To achieve adequate ionization and fragmentation for most general purposes the ionization is carried out[80] with electrons possessing an energy of 70 eV. Operation at very low energy (e.g. about 20 eV) produces spectra showing little more than the molecular peak (*see* **Parent peak**) and the **base peak**.

Electron energy analyser (pe.). Any device employed to measure the energy distribution of emitted electrons. Most analysers are based upon deflection of electrons in electrostatic or magnetic fields or their behaviours at potential barriers. Such instruments must achieve resolving powers of 5000 (i.e. 0.2 eV at 1000 eV).

Early electron analysers were based upon magnetic deflection, but electrostatic instruments are now preferred as they are more easily screened from external magnetic fields. The electrons follow curved paths under the influence of the electrostatic field and are brought to a focus on the collector cup which is linked to an electron multiplier.[81,82] The various electrons are resolved according to their kinetic energies and brought to a focus by varying the applied electrostatic potentials.

Retarding field energy analysers, by comparison, really serve as energy filters as the kinetic energy of the electrons is established by determining the magnitude of the potential barrier they can overcome. For this purpose the electrons can only reach the collector by passing through a retarding grid to which a scanning potential is applied.[83,84] Scanning of the spectrum is carried out by progressively reducing the retarding potential and recording the collector current.

Electron energy loss spectroscopy (EELS). (pe.). *See* **Electron impact spectroscopy**.

Electron gun (pe.). Production of beams of electrons for Auger electron spectroscopy (*see* **Auger effect**) may be achieved by using a thermionic **cathode** source in the form of an electron gun. Such guns are modified forms of the electron beam generators used in oscilloscope tubes. The cathode consists of a thoria coated tungsten filament operating at about 1700–1800 K and the beam is accelerated along a focusing tube by a voltage of 2–5 eV. Final adjustment of the beam on the sample is achieved by two pairs of electrostatic deflectors.[85]

It has, however, been claimed that the use of electron beams over X-rays for this purpose is limited.[86] The electron gun is probably of more value in its role as a producer of X-rays when the electron beam is directed onto an appropriate **anode**.

Electronic spectra (a.s., em., u.v.). The promotion of an electron from a lower energy state to a higher energy state is brought about by the absorption of energy by the atomic or molecular species. This energy may be measured in the form of atomic absorption spectra (*see* **Atomic absorption spectroscopy**) or as an **ultraviolet/visible spectrum**. Where an electron drops from a higher energy state to a lower energy state there is a loss of energy from the species which may be measured in the form of an emission spectrum (*see* **Emission spectrometry**).

The emission spectra from atoms occur as a series of discrete lines corresponding to fixed wavelengths (*see* **Atomic spectra**), but the absorption spectra for molecules are normally broad bands as the electronic transitions are accompanied by vibrational and rotational transitions (*see* **Bands (ultraviolet)**; **Ultraviolet/visible spectrum**).

The energy differences between electronic levels vary from < 200 kJ mole^{-1} in visible spectra to > 1200 kJ mole^{-1} in the far ultraviolet.

Electron impact induced reactions (m.s.). A term used to refer to the rearrangement of **fragment ions** produced as a result of electron bombardment of molecules in mass spectrometers. *See also* **McLafferty rearrangement**.

Electron impact spectroscopy (Electron energy loss spectroscopy (EELS)) (pe.). An inclusive term for forms of **photoelectron spectroscopy** in which electron beams are used to induce transitions between electronic energy levels for the study of excited molecular states.

Electron ionization (EI) (m.s.). The process of forming positive ions and fragmenting molecules by bombarding compounds in a mass spectrometer with electrons using an **electron bombardment ion source.**

Electron multiplier. *See* **Photomultiplier.**

Electron nuclear double resonance (e.s.r.). *See* **ENDOR.**

Electron paramagnetic resonance. *See* **Electron spin resonance.**

Electron spectroscopy for chemical analysis (pe.). *See* **ESCA.**

Electron spin resonance (Electron paramagnetic resonance). The study of atoms, ions, free radicals and molecules possessing an odd number of electrons or unpaired electrons which display characteristic magnetic properties.[87] These arise from the spin properties of the unpaired electrons. Under the influence of a magnetic field the magnetic moment of the spinning electron may occupy either of two possible orientations, aligned or opposed to the magnetic field, corresponding to the two energy levels for the electron spin. The energy levels have a difference of

$$\Delta E = h\nu = g\mu_B B$$

At resonance the absorption of energy causes the magnetic moment of the spinning electron to rise (flip) from the lower to the higher energy state.

The e.s.r. spectra are complicated by the fact that coupling occurs between the spinning electron and adjacent spinning nuclei, so that the absorption spectrum (usually recorded as a derivative curve (*see* **Derivative spectra**)), consists of a multiplicity determined by the number, nature and position of coupling nuclei. It is this feature of the spectra that makes them of value in structural elucidation. Typical values for coupling constants are between 2-30 gauss, and n non-equivalent protons produce 2^n hyperfine lines in the e.s.r. spectrum.

Modern e.s.r. spectrometers can detect free radicals at concentrations as low as 10^{-13} mole. In a field of about 3400 gauss the frequency required for resonance is 9.5 × 10⁹ Hz. and investigations have been carried out in a wide range of chemical and biochemical systems.[88]

Figure 18 here

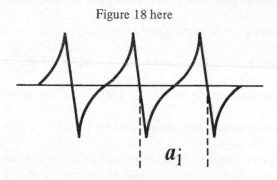

Figure 18. Splitting of an e.s.r. signal by interaction between an unpaired electron and an atomic nucleus possessing a spin quantum number $I = 1$. The measurement a_i is the **coupling constant**.

Electronvolt (eV). A unit of energy traditionally associated with the motion of atomic and molecular particles. It is not a recommended **SI unit** although accepted for use in specialized areas. The conversion factor for the unit is

$$1 \, eV \approx 1.6022 \times 10^{-19} \, J$$

Electrothermal atomizer (a.s.). An inclusive term now used for the wide range of flameless atomizers, usually with carbon rods or filaments, which are electrically heated to high temperatures to cause volatilization and atomization of samples for atomic spectroscopy. Such systems are now widely used in studies in both **atomic absorption spectroscopy** and atomic **emission spectrometry** and it has been shown that the sensitivity using such carbon furnaces may be a hundred times that employing a flame.[89] One major factor greatly affecting electrothermal atomization is the rate of heating of the sample.[90] A disadvantage of using these systems is that the signal from the sample may only last for a few seconds, whereas with nebulization of solutions into flames a signal is obtained for as long as the solution is sprayed. *See also* **Carbon filament atom reservoir**; **Delves cup**; **Massmann furnace**; **Tantalum boat**.

Emission spectrometry (em.). Electrons in the ground state may be promoted to higher energy levels by excitation of the atoms using electrical or thermal means. On reverting to the ground state, or any intermediates energy level, the absorbed energy is released and may be measured and recorded to give a representation of the emission spectrum for the substance.

Three forms of emission spectra exist: these are **continuous spectra** as produced by incandescent solids, band spectra as given by molecules, and **line spectra** obtained from atoms.

One form of emission spectrometry that is extensively used for atomic spectra is that in which the characteristic lines for metals are obtained by placing alloys between electrodes across which an arc or spark is discharged and recording the results on a photographic plate. This is the traditional way to obtain a qualitative analysis of complex metal alloys.

Emission spectrometry using flames and inductively coupled plasmas is more suitable for trace and quantitative analysis and employs instrumental procedures similar to that for atomic absorption spectrometry. Detection limits are of the order of 0.1-10 ppm for many metals and may be even lower if the **inductive coupled plasma** is used.[91]

ENDOR (Electron nuclear double resonance) (e.s.r.). Improved resolution of e.s.r. spectra is obtained by the ENDOR technique in which the e.s.r. line is broadened and hyperfine details enhanced.[92] This is achieved by irradiating the sample with nuclear resonance frequencies while observing the e.s.r. absorption signal.[93] *See also* **Overhauser effect**.

Energy level. The total energy of an atom consists of the sum of the energies of its constituent sub-atomic particles. Each of these may exist in one of a number of energy states, and energy is absorbed or emitted by atoms as a result of transitions between the energy levels representing these states.

Transitions between electron energy levels in atoms give rise to **atomic spectra** while similar transitions in the energy levels of bonding electrons in molecules produce **molecular spectra** studied in the ultraviolet and visible regions.

Transitions between nuclear spin energy levels give rise to nuclear magnetic resonance spectra (*see* **Nuclear magnetic resonance spectroscopy**).

A diagrammatic representation of the transitions between energy levels is shown when considering the **Franck–Condon principle**.

Enrichers. A collective term that has been used to refer to the wide range of systems used for interfacing gas chromatography and liquid chromatography with mass spectrometers and infrared spectrometers. *See also* **GC/IR interfaces; GC/MS interfaces; LC/IR interfaces; LC/MC interfaces.**

EPA mixed solvent (pl.). Studies of phosphorescence carried out in a solid matrix at low temperatures necessitate the use of a clear medium. The EPA mixed solvent is particularly suitable for this purpose, consisting of diethyl ether, 2-methylbutane and ethanol in the ratio 5:5:2 by volume. It solidifies to a clear glass when frozen in liquid nitrogen. *See also* **Shpol'skii effect.**

ESCA (pe.). An acronym for 'electron spectroscopy for chemical analysis' introduced[94] in 1967. It refers to X-ray induced photoelectron spectroscopy particularly applied to chemical analysis with special emphasis on studies of surface layers, catalysts and protective coatings.[95]

Escape depth (pe.). *See* **Inelastic mean free path.**

e.s.r. *See* **Electron spin resonance.**

Excimer (pl.). The term (from excited dimer) was originally introduced for what were believed to be long-lived species exhibiting fluorescence after an excited molecule had reacted with a second molecule.[96] Other explanations have been found for the phenomena in this classification[97] and the word 'excimer' is now used for short-lived excited dimers of the type obtained with pyrene in which the dimer fluorescence spectrum becomes more pronounced with increased concentration of the fluorescer in solution.

Excited state. An atom in which one or more electrons have been promoted from the lowest energy level to higher energy levels is said to be in excited state. More generally the expression is used to refer to the energy state of any molecule, atom or sub-atomic particle above that of the highest occupied energy level in the ground state. *See also* **Ground state.**

External reference (n.m.r.). The reference standard in n.m.r. may be contained in a capillary tube within the n.m.r. sample tube. In this form it constitutes an external standard and because of the glass walls of the capillary tube does not experience exactly the same magnetic effects as does the sample under study. This can lead to slight differences in value for **chemical shifts** compared with results using **internal references**. As the normal reference material **TMS** is insoluble in solutions made up in D_2O it is common to employ external references, although an alternative to this problem is to use the water-soluble secondary standard **DSS**.

Extinction. *See* **Absorbance.**

Extinction coefficient (E) (u.v.). Traditionally absorbance obtained for a 1% solution of material in a cell of 1 cm path length and is normally used for substances of unknown molecular weight. In this case it is written in the form $E_{1\,cm}^{1\%}$. The relationship between this and the **molar absorption coefficient** is given by

$$E_{1\,cm}^{1\%} = \frac{10\epsilon}{\text{molecular weight}}$$

It has been suggested that the word 'absorptivity' should be used in place of extinction coefficient, but this has not met with general acceptance. Currently the recommendation is that the expression employed internationally should be 'absorption coefficient' using the symbol a such that

$$a = \frac{A}{l}$$

F

Faraday cup (m.s.). The simplest form of electrical detector, consisting of a metal cylinder closed at one end which serves as the collecting electrode. The ion beam travels along the axis of the cylinder and impinges at a point on the bottom producing a small d.c. voltage which is fed to an amplifier and recorder. This type of detector has been of particular value in isotopic abundance studies in mass spectrometry although it is not as sensitive as a **photomultiplier**. In most mass spectrometers the Faraday cup assembly is fitted with other electrodes which serve as ion suppressors to remove scattered ions.

Faraday effect. Faraday discovered[98] that if polarized light is passed through any liquid or solution placed in a magnetic field the plane of polarization is rotated. This is known as the Faraday effect and is due to the influence of the applied field upon the movement of the electrons. For the majority of compounds the rotation is in the same direction as the current producing the magnetic field.

Far-infrared region (i.r.). *See* **Infrared spectra**.

Fermi contact interaction (e.s.r.). All *s* electrons and those in hybrid orbitals containing some *s* character have a probability of being at the centre of an atom and in such as position would be subject to a different magnetic force from that experienced further from the centre. As it approaches the nucleus the electron acquires a large kinetic energy and its velocity approaches that of light. Calculations for this state of affairs were carried out by Fermi and it has become known as the Fermi contact interaction.

This interaction involves coupling between the nuclear and electron magnetic moments and leads to differences in the hyperfine splitting of the e.s.r. signal from unpaired electrons in σ molecular orbitals and those possessing some σ character.

Fermi energy (E_F) **and Fermi level** (pe.). The Fermi energy represents the energy of electrons in a conducting metal at the Fermi level and is given by the equation:

$$E_F = \frac{1}{2m} \left(\frac{3h^3 N}{8} \right)^{2/3}$$

Where N is the number of electrons in unit volume of real space in which all energy levels are occupied up to E_F.

Fermi resonance (i.r.). Accidental **degeneracy** arising in polyatomic molecules in which two different vibrational states accidentally possess approximately the same energy and interact with each other. The result of this resonance is that any difference between the vibrational levels is increased. It is particularly common between a fundamental band and an overtone or combination band.

Féry glass prism (u.v.). Use of a collimating mirror in monochromation can be obviated by using the Féry back reflecting prism with curved faces one of which is coated with an aluminium film to serve as a reflecting surface.[99] The prism was originally developed to use in the Féry spectrograph which acts as collimator and objective in addition to carrying out dispersion.[100] This is brought about as a result of the prism curvature which is such that rays diverging from a point on the circumference of a circle are reflected back and focused at another point on the circle.

Field-free region (m.s.). In magnetic deflection mass spectrometers employing a magnetic analyser of less than 180° there is a small region in which the positive ions are not acted upon by the magnetic field. This is a section between the **accelerating slits** and the magnetic field; it is known as the field-free region and arises owing to the geometry of the instrument. In 180° magnetic analyser instruments it is avoided by placing the source and accelerator slits within the complete magnetic field.

In **double-focusing mass spectrometers** a second field-free region exists between the electrostatic field and the magnetic field. In some instances specially large field-free regions are built into the design to assist in the study of **metastable ions.**[101]

Field desorption (FD) (Surface ionization) (m.s.). A method of ionization for mass spectrometry applied to readily volatile solids with low ionization potentials. The compounds are coated on a tungsten knife edge or filament anode which is then heated to 2000 °C whilst subject to **field ionization** conditions. Molecules evaporated from the surface tend to lose electrons to give position **molecular ions** but greater fragmentation occurs than in the field ionization technique.

Field ionization (FI) (m.s.). A relatively gentle ionization procedure in mass spectrometry in which the sample is subjected to a high voltage field of about 10 kV using a knife edge or wire anode placed about 1 mm away from the cathode forming the exist slit of the ionization chamber. The high voltage leads to loss of electrons by the molecules, producing a high yield of molecular ions and little fragmentation so that determination of molecular weights and empirical formulas are obviated. *See also* **Chemical ionization**; **Electron ionization**; **Field desorption.**

Filters. The simplest form of wavelength selector, used extensively in colorimeters and simple visible spectrophotometers. Specially selected filters are employed in fluorimeters in order to give a narrow irradiation band and to isolate the fluorescent wavelength.

A filter is a semi-transparent material, usually in the form of a thin glass plate, which is capable of absorbing unwanted electromagnetic radiation and transmitting a narrow band of desired wavelengths. In order to cover a range of wavelengths and give selective transmission a series of different filters is required. The selectivity is achieved by adding a small amount of special materials, such as rare earths, to the glass formulation. High quality filters usually have **effective bandwidths** of up to 10 nm and transmittances can be as high as 50 per cent. *See also* **Bandpass filter**; **Interference filter**; **Wratten filter.**

Fingerprint region (i.r.). The 7–11 μm (1428–909 cm^{-1}) portion of the infrared spectrum of a substance usually contains a large number of absorption peaks and shows fairly clear differences between even closely related compounds. Although not every absorption peak can always be assigned to particular vibrations in a molecule, the fact that each spectrum is characteristic for the substance from which it has been obtained enables identification to be made by using the spectra in this region as 'fingerprints' and looking for the identical absorption pattern in a library of spectra.[102]

First-order spin pattern (n.m.r.). Any two coupling groups of protons for which the difference between the **chemical shifts**, $\Delta\nu$, is at least six times the coupling constant J (i.e. $\Delta\nu > 6J$). Under these conditions the ratios of peak heights in the absorption multiplets are predictable from the **Pascal triangle**. In other cases in which $\Delta\nu \gtrless 6J$ the simple multiplet ratios apply less and less as $\Delta\nu$ decreases with a corresponding trend away from a first-order pattern. *See also* **Multiplets**.

Fishhook symbol (m.s.). A symbol introduced[103] to denote the movement of single electrons in molecular and fragment ions as an aid to explaining rearrangement reactions that occur in the mass spectrometer. The full arrow is used to show the transfer of two electrons. The use of both the fishhook and the full arrow are illustrated in the following reaction sequence:

Fixation pair (em.). In the use of **internal standards** in quantitative emission spectroscopy it is usual to compare an emission line of the standard with one from the unknown. If the two lines can only be used under fixed (defined) excitation conditions they are known as a fixation pair. As they respond differently to any change in the excitation a fixation pair can be used to detect any variations in exposure. However, because of the difficulties in maintaining completely reproducible conditions for quantitative studies, it is more common to use an **homologous pair**.

Fixed path-length cell. *See* **Liquid cell.**

Flameless atomic absorption (a.s.). *See* **Carbon filament atom reservoir**; **Electrothermal atomizers**; **Massmann furnace.**

Flame photometer (a.s.). An instrument used to study the characteristic radiation obtained from elements when they are introduced into a flame. The intensity of the emission for any particular element is proportional to the concentration.

The flame photometers consists of (*a*) the nebulizer, (*b*) the burner, (*c*) the optical system, (*d*) the detector and (*e*) the recorder. In simple flame photometers the monochromation for separate elements is achieved in each case by using a specially selected filter,[104] in more sophisticated instruments a grating is used for monochromation.

The success of flame photometry is greatly dependent upon the flame temperature and the correct filter. It is not usually considered as sensitive a method of analysis as is **atomic absorption spectroscopy** but this contention has been questioned[105] and in some respects its sensitivity is superior. *See also* **Emission spectrometry.**

Flame source (a.s.). Both atomic absorption spectroscopy and flame photometry use flames as a common means of exciting samples for spectroscopic studies.[106] Flame temperatures between 2000 and 3000 °C are required for most elements and are attained by burning mixtures of gases such as air and methane or air and acetylene. The sample may be sprayed into the flame or heated in a small crucible (*see* **Delves cup**). Flame geometry is of considerable importance in achieving maximum sensitivity, the optimum position generally being in the higher, wider part of the flame.[107]

Attempts have been made to develop a general-purpose gas mixture to obviate the necessity of changing burners and gases to determine different elements. Propane–butane–nitrous oxide and acetylene–nitrous oxide–air are widely applicable and come near to being the ideal mixtures.[108] *See also* **Fuel gases.**

Fluorescence (pl.). Fluorescence takes place when an electron in an **excited state** has failed to lose its excess energy by collisions with other molecules and so returns to a vibrational level in the ground state by the direct release of excess energy. This occurs after a **singlet state** lifetime of about 10^{-9} s and takes place from the lowest vibrational level of the first excited singlet state. The wavelength of fluorescence spectral bands is theoretically independent of the excitation wavelength although the absolute intensity of fluorescence is dependent on this. Nevertheless, for a substance to exhibit fluorescence the exciting radiation must be of a wavelength that the substance will absorb.[109]

The fluorescence spectrum appears on the long-wavelength side of the original absorption band and is nearly always associated with the electron system of an unsaturated molecule. The presence of certain chemicals can lead to **quenching** of the fluorescence. *See also* **Triplet state.**

Fluorescence quantum counter (pl.). As the fluorescent response of many substances in solution is independent of the wavelength of the incident radiation it is possible to use such fluorophors to measure the quantum intensity of the radiation over a wide spectrum range. The fluorescence quantum counter consists of a solution of the fluorescing material in a detector cell with a path length great enough to absorb all the incident light after it has penetrated only a short path of the total cell path. Any fluorescence produced then has to pass through the remainder of the solution path length. By this means the solution serves as a filter and only allows the fluorescent wavelength to reach the photocell. The amount of fluorescence is thus a measure of the quantity of incident radiation.[110]

Fluorolube (i.r.). A mixture of fluorinated hydrocarbons used as a mulling agent for solids in infrared spectroscopy. It is specially used to obtain the spectrum in those regions in which **Nujol** absorption bands appear. By running spectra of **mulls** prepared with the two individual mulling agents it is possible to obtain the complete spectrum of the sample.

Fluorometer (pl.). It has been suggested[111] that this name should be used for instruments designed for measuring the lifetime of fluorescent species. In the UK the word fluorimeter is more often employed. *See also* **Spectrofluorimeter**.

Fluorophor (pl). Any molecule in an excited state that is capable of exhibiting **fluorescence**.

Force constant (i.r.). The restoring force for a unit displacement from an equilibrium position for a system of two point masses (m_1 and m_2). It is employed in the following expression to obtain values for infrared absorption maxima

$$\tilde{\nu} = \frac{1}{2\pi c}\left(\frac{f}{\dfrac{m_1 m_2}{m_1 + m_2}}\right)^{1/2} \text{cm}^{-1}$$

An approximate value for f is obtained from the expression[112]

$$f = aN\left\{\frac{\chi_1 \chi_2}{d^2}\right\}^{3/4} + b$$

in which N is the bond order (number of bonds between the two atoms), χ_1 and χ_2 are electronegativities of the atoms, d is the internuclear distance, and a and b constants with values 1.67 and 0.30 respectively.

For a single bond $f \approx 5 \times 10^{-5}$ dyn cm^{-1} = 5 × 10^{-9} N m^{-1}

For a double bond $f \approx 10 \times 10^{-5}$ dyn cm^{-1} = 10 × 10^{-9} N m^{-1}

For a triple bond $f \approx 15 \times 10^{-5}$ dyn cm^{-1} = 15 × 10^{-9} N m^{-1}

Fourier-transform infrared. *See* **Fourier-transform spectroscopy.**

Fourier-transform n.m.r. A method of increasing the sensitivity of n.m.r. spectroscopy first proposed by Ernst and Anderson.[113] *See* **Fourier-transform spectroscopy.**

Fourier-transform spectroscopy. The application of Fourier analysis to spectroscopy has led to the development of special instruments designed to give improved spectra. Such instruments study the whole of the selected spectral range all the time in order to give an improved signal to noise ratio compared to traditional instruments employing progressive scanning.

In far-infrared spectroscopy the Fourier system employs two light beams to provide an interferogram[114,115] which is presented as a digital signal and recorded on punch tape. The final spectrum is obtained from a computer read-out. Some modern instruments possess built-in computers to process the data back directly into a convenient form. It has been found to be of special value in obtaining infrared spectra of surfaces[116] in preference to **attenuated total reflectance** methods and enables a complete spectrum[117] to be recorded in 0.25 s. The technique has also been applied to n.m.r. spectroscopy[118] by subjecting the sample to a series of r.f. pulses and Fourier transforming the free induction decay signals obtained. Less sensitive nuclides including ^{31}P, ^{14}N and ^{13}C have been studied by this technique as sensitivities are increased by factors of between 50–80.

Fragment ions (m.s.). Cleavage of bonds in molecular ions in the **mass spectrometer** leads to the formation of ions from portions of the original molecule. Sometimes these fragments undergo rearrangements before being detected and recorded on the spectrum. Fragment ions are shown as mass/charge peaks lower than the parent molecular peak on the spectrum. Their value is in the indication they give to the nature of groups and substituents in the parent molecule. *See* **Bar graph.**

Franck–Condon principle. Electronic transitions tend to take place from vibrational levels of one electronic state to vibrational levels of other electronic states for which the internuclear distances are the same. These transitions occur at the points at which the kinetic energies of the nuclei are at a minimum (i.e. at the limiting positions of the vibrations). Thus during the transitions there is no appreciable change in internuclear distance or momentum.[119]

On a potential energy diagram the change is represented by a vertical line from a vibrational level in one electronic state to a vibrational level in the other electronic state. Absorption of energy is shown by an upward arrow and emission by a downward arrow.

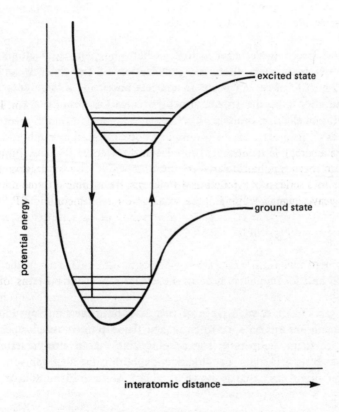

Figure 19. Franck–Condon principle.

Freedom, degrees of (i.r.). *See* **Degrees of freedom**.

Free radicals (e.s.r.). Free radicals come within the classification of systems possessing a net electron spin angular momentum that will give an **electron spin resonance** spectrum. Free radicals are molecules possessing an odd electron, frequently produced by homolytic fission of a chemical bond

$$A:X \longrightarrow A^{\cdot} + X.$$

   molecule   free radicals

Most free radicals occur as intermediates in chemical reactions and have relatively short lifetimes; stabilization of free radicals and an increase in lifetime can occur in radicals possessing a large degree of unsaturation (e.g. the triphenylmethyl radical) as a result of the combination of electron orbitals.

Frequency (ν). Frequency is used with reference to any periodic phenomenon and particularly to the number of cycles per unit time and varies inversely with the wavelength. It is proportional to the energy of a quantum, $E(\phi)$.

$$\nu = \frac{c}{\lambda} = \frac{E(\phi)}{h}$$

The **SI unit** for frequency is hertz (i.e. cycles per second = **hertz**).

Fuel gases (a.s.). A wide range of fuel gases have been employed[120],[121] in atomic absorption spectroscopy and flame photometry, usually to produce flame temperatures exceeding 2000 °C. In most cases these temperatures can only be obtained by buring the fuel gas with air, oxygen or nitrous oxide as shown in table 5. *See also* **Flame source**.

Table 5

	Fuel gas	with air	Temperature/°C with oxygen	with nitrous oxide
town gas	(mainly methane)	1800	2700	
	propane	1950	2800	2850
	butane	1900	2900	
	hydrogen	2050	2700	2850
	acetylene	2400	3100	2950
	cyanogen	2330	4500	

G

Gallium cut-offs (m.s.). One method of introducing a liquid sample with a low vapour pressure ($<$ 0.1 torr) into a mass spectrometer is by passing a pipette containing the liquid through a molten metal seal covering a sintered glass disc leading to the reservoir of the sampling unit. The low pressure in the reservoir causes the liquid to flow through the sintered disc. Molten gallium has become the standard material employed for this purpose as it is liquid over the range 30–1983 °C and operating temperature, commonly up to 400 °C, are beyond the limits of other inlet system seals.[122] Mercury is unsuitable for this purpose because of its high vapour pressure.

Gas cells (i.r.). To obtain an adequate absorption of infrared radiation by gases and vapours a longer sample path length is required than for liquids and solids. Because of this normal gas cells are about 10 cm long consisting of a glass tube with infrared transparent windows at the ends. For substances possessing low vapour pressures, or available in only small amounts special extended path-length cells are employed. These reflect the infrared radiation back and forth through the vapour by means of a mirror system within the cell and an effective path length of 10 metres or more can be obtained (see Figure 20, page 75).

Gaussian curve (distribution). The most common frequency distribution for continuous variables is the Gaussian (or normal) distribution; the equation describing this gives the probability at a value x for a set of variables possessing a mean μ and a standard deviation σ.

$$P\{x\} = \frac{1}{(2\pi)^{1/2}\sigma} \exp \left\{ -\frac{(x - \mu)^2}{2\sigma^2} \right\}$$

The Gaussian curve shape is such that the half-width $W/2$ at half-height $H/2$ are related by the equation

$$\frac{W}{2} = \frac{1.476}{H}$$

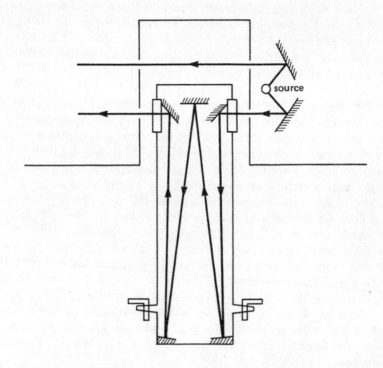

Figure 20. Extended path-length gas cell

Infrared absorption curves correspond to Gaussian shapes, which
are symmetrical if recorded on a linear wavenumber abscissa and
asymmetric if recorded with a linear wavelength scale. *See also* **Lorenz-
tian curve**.

GC/IR interfaces. Problems associated with GC/IR interfacing[123] have
been greatly overcome by the use of **Fourier-transform** techniques such
that detection limits in the nanogram range are now possible.[124] In such
systems the eluate from the gas chromatograph is passed to a flow
splitter one part going to the GC detector which triggers the Fourier

interferometer when a peak is detected, and the other passing through a 40 cm × 0.3 cm i.d. gold-coated glass tube fitted with KBr windows at the ends. Condensation of eluate is prevented by heating the light tube along its length. To obviate mixing of separate compounds laminar flow must be maintained through the light tube. Some cells have been constructed with effective volumes of less than 1 cm^3.

GC/MS interfaces. To obtain a suitable concentration of sample from a gas chromatograph that can be led directly into a mass spectrometer it is necessary to remove the majority of the carrier gas. Enrichment has been achieved by a variety of interfaces.[125] Early interfacing was carried out by direct linking of the gas chromatograph to a **time-of-flight mass spectrometer** through either a control valve or effluent splitter. Later systems have either removed the carrier gas by preferential diffusion through a sintered glass tube or have removed the gas by pumping it away as it passes between jets, as shown in Figure 21. Another approach has been the use of a permeable silicone rubber septum which allows passage of organic molecules but not of the carrier gas. This gives sample recoveries > 50%, but suffers from a small time lag. More recent approaches have been with direct coupling in which GC eluates are passed through an open ended capillary from which the carrier gas is pumped away whilst the solute passes to the mass spectrometer ionization chamber. *See also* **Watson–Biemann separator.**

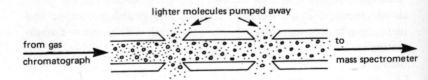

Figure 21. GC/MS jet separator

Geissler tubes (pl., u.v.). Special gas discharge tubes constructed to concentrate the discharge in a capillary tube joining two glass bulbs containing the anode and cathode respectively. The tubes are operated by induction cells and produce intense glow discharges when small quantities of gases or vapours are contained inside the tube.

Getter-ion pump. *See* **Sputter-ion pump.**

g **factor** (e.s.r., n.m.r.). *See* **Landé** *g* **factor.**

Globar (i.r.). A distribution of infrared radiation similar to that obtained from a black body (*see* **black body radiation**) is produced from the globar.[126] It is made from sintered silicon carbide shaped into a rod or cylinder about 3 cm long and 6 mm in diameter and is one of the four sources of radiation employed in infrared spectrophotometers. The rod is electrically heated between 750 and 1500 °C operating at low voltage and high current; no external preheating is required before it can be used. *See also* **Nernst glower**; **Nichrome source**; **Opperman source.**

Golay coils (n.m.r.). *See* **Shim coils.**

Golay detector (i.r.). *See* **Pneumatic cell.**

Graphite rod (a.s.). *See* **Carbon filament atom reservoir.**

Gratings. The conventional diffraction grating consists of a series of closely spaced parallel lines scored onto the surface of a thin glass plate. Spectra are produced by **diffraction** of the wavefront at the grating surface and the angle (δ) at which these are observed is given by the equation

$$n\lambda = a(\sin \delta \pm \sin i)$$

where n is the order of the spectrum, λ the wavelength, a the distance between adjacent grating lines and i the angle of the incident radiation.

In practice confusion between the first ($n = 1$) and higher orders of spectra is prevented by removing the higher orders by use of a filter or prism in the light path. Most modern diffraction gratings for spectroscopy are echelette reflection gratings consisting of a series of fine angled ridges ruled on a place surface. **Dispersion** from gratings is more uniform than is that from prisms, and consequently grating spectrophotometers can maintain high resolution over a longer wavelength range.

For u.v./visible spectrophotometers the gratings used have between 10 000 and 30 000 lines cm^{-1}, whilst in the infrared the range is 100 to 3000 lines cm^{-1}. Full coverage of the middle-infrared region is frequently obtained by using two gratings, the first to cover the range 4000-1200 cm^{-1} (2.5-8.33 μm) and the other for the 1333-400 cm^{-1} (7.5-2.5 μm) range. Some recent instruments now cover this portion of the infrared with a single grating.

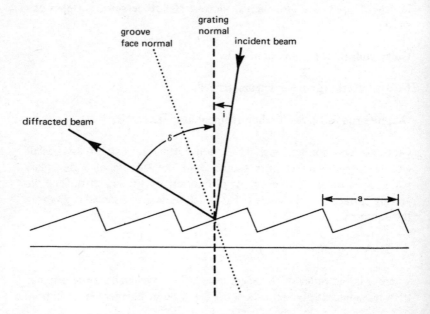

Figure 22. Diffraction grating

Ground state. A general expression used to refer to the lowest energy or unexcited state of a system. Molecules, for example, possess a ground rotational state as well as ground vibrational and ground electronic state. *See also* **Excited state** and **Franck–Condon principle**.

Gyromagnetic ratio (n.m.r.). *See* **Magnetogyric ratio**.

H

Half-wave plates (r.s.). Special types of **retardation plates** used to create half a **wavelength** or 180° phase difference between the phases of the ordinary and extraordinary components of the polarized light.

Hard X-rays. *See* **X-rays.**

Harmonic vibrations (i.r.). *See* **Overtones.**

h **bar.** *See* **Planck constant.**

Heated graphite atomizer (a.s.). *See* **Carbon filament atom reservoir; Massmann furnace.**

Heated inlet systems (m.s.). Solids and liquids possessing low vapour pressures can be introduced into the mass spectrophotometer by several different heated inlet systems. In some cases the **batch inlet sampling system** can be used if it is fully heated along its length to the ionization chamber. A **direct insertion probe** operating at elevated temperatures is frequently used for solids, whilst liquids can be introduced through a molten **gallium cut-off.**

Helmholtz pair (n.m.r.). *See* **Sweep coils.**

Henke gun (pe.). One of the early sources for the production of X-rays for photoelectron spectroscopy was the Henke gun,[127] and this has served as the basis for the development of other devices. Its main features are a **cathode** filament positioned out of direct line with the **anode** which is maintained at a positive potential of about 3 kV. Any scattered electrons are drawn towards the anode and do not impinge on the window through which the **X-rays** are emitted.

Hertz (Hz) (n.m.r.). An **SI unit** for frequency which has now superseded the former expression cycles per second (cs^{-1}). It is used for the number of repetitions of a regular occurrence in one second and is expressed in reciprocal sounds (s^{-1}).

Heterolytic cleavage (m.s.). Fragmentations of positive ions by transfer of a pair of electrons to one of the fragments. It is represented in mechanistic equations by a full arrow. *See also* **Fishhook symbols**; **Homolytic cleavage**.

$$RCH_2Cl \xrightarrow{\;-e\;} RCH_2 \overset{+\bullet}{:\underset{\bullet\bullet}{Cl}} : \;\longrightarrow\; RCH_2^+ + :\underset{\bullet\bullet}{\overset{\bullet\bullet}{Cl}} :$$

Heteronuclear double resonance (n.m.r.). **Spin–spin decoupling** techniques can be applied to identical nuclei (homonuclear) or to different ones (heteronuclear), so that the **multiple resonance** methods can be used to study the multiplicities arising from the coupling between such dissimilar elements as hydrogen and phosphorus.

Heteronuclear double resonance can obviously be carried out in two ways, either of the two elements being irradiated. The representation $^1H-\{^{31}P\}$ indicates that in studying the proton magnetic spectrum of a substance it is the ^{31}P nucleus that has been decoupled. *See also* **Homonuclear double resonance**.

Higher-order spin patterns (n.m.r.). *See* **Second-order spin patterns**.

High-intensity lamps (a.s.). It has been shown that radiation emitted by **hollow cathode lamps** results from a two-stage process involving sputtering followed by excitation. In the high-intensity lamps two extra electrodes are built in to produce an additional current across the hollow cathode.[128,129] This assists preferential excitation of the sputtered atoms and has enabled increased sensitivity to be obtained in the determination of some elements.

Hohlraum (i.r.). A laboratory device which produces **black-body radiation**, usually as a standard for comparison with infrared radiation sources. It consists of a closed metal tube, blackened on the inside, with a narrow slit cut into one of the flat ends. On heating the tube the radiation escaping from the slit is virtually identical with that expected from a theoretically black body. *See also* **Globar**; **Nernst glower**; **Nichrome source**; **Opperman source**.

Hollow cathode lamp (a.s.). Despite the fact that **electrodeless discharge tubes** for some elements are now available the hollow cathode lamp is still the main type of source used for atomic absorption studies up to now.[130,131,132] The lamp is chosen such that the emitting cathode is of the same element as that being studied in the flame. The cathode is in the shape of a cylinder alongside the anode and is enclosed in a borosilicate or quartz housing containing an inert fillter gas, neon or argon, at about 2 mm pressure. The application of a high potential of about 800 V causes a discharge which creates ions of the inert gas. These are accelerated to the cathode and on collision excite the cathode element to emission.[133] Multi-element lamps are also available in which the cathodes are made from alloys or sintered powders.

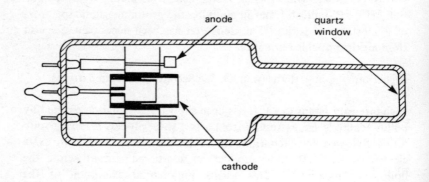

Figure 23. Hollow cathode lamp

Homogeneity (n.m.r.). When applied to magnets this refers to the uniformity of the magnetic field around the sample. For high-resolution n.m.r. the magnetic field in this region should not vary by more than

1 part in 10^8. For low-resolution **broad-band n.m.r.** the acceptable variation is 1 part in 10^5.

Homologous pair (em.). In the use of emission spectroscopy for quantitative analysis by the **internal standard** method it is usual to overcome variations in time of exposure and excitation conditions by comparing a line of the unknown with a line of the internal standard possessing roughly the same intensity. These two lines must be such that they are affected in the same way by any changes in the excitation conditions. This means that the ratio of the intensities of these two lines should be constant even if there are variations in exposure or development. *See also* **Fixation pair.**

Homolytic cleavage (m.s.). Fragmentation of ions in mass spectrometry frequently involves single electron transfers to each of the two resulting fragments. This is termed homolytic cleavage and the single electron movements are indicated by **fishhook symbols** as shown in the equation below. (*See also* **Heterolytic cleavage.**)

Homonuclear double resonance (n.m.r.). Spin–spin decoupling applied to identical nuclei can be carried out using **multiple resonance** techniques. Under these conditions when the nuclei are isotopically identical it is called homonuclear double resonance. It is represented by $^1H-\{^1H\}$ indicating that the proton within the brackets is the one that has been irradiated and therefore decoupled. *See also* **Heteronuclear double resonance.**

Hooke law (i.r., r.s.). In its normal form this law states that an elastic body suffers a strain that is proportional to the stress imposed upon it. In more general terms this can be expressed to mean that the elongation of an elastic object will be proportional to the force applied to it.

The law is equally applicable to the vibrational nature of molecular bonds and under these conditions it is possible to employ an equation relating the frequency of the oscillation, the force constant of the bond and the masses of the two atoms

$$\nu = \frac{1}{2\pi} \left(\frac{f}{\frac{m_1 m_2}{m_1 + m_2}} \right)^{1/2}$$

Hybrid T (Magic T) (e.s.r.). An arrangement of tubular **waveguides** used in e.s.r. spectrometers to prevent the microwave power from the **Klystron oscillator** from passing straight to the crystal detector.

Energy directed from the klystron to arm A is split equally in two directions through the arms B and C to the balancing load and resonant cavity respectively. No power reaches arm D, the crystal detector, if B and C are balanced. But when an e.s.r. absorption occurs in the cavity the system is out of balance and this results in a signal from C which passes to the detector in D.

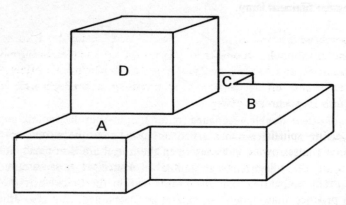

Figure 24. Hybrid T

Hydrogen bonds. Hydrogen atoms attached to atoms possessing unpaired electrons, such as oxygen and nitrogen, are able to undergo a weak form of association with groups possessing easily polarized electrons. The hydrogen atom, whilst being held by its former atom is attracted to the second group and in molecular structures this is represented by a dotted line between the two atoms (e.g. $-OH \ldots : O = C <$).

These associations may occur between atoms in the same molecule, intramolecular hydrogen bonding, or between atoms in different molecules, intermolecular hydrogen bonding. The existence of hydrogen bonding gives rise to characteristic absorptions in spectra; it is particularly noticed in infrared spectra[134] for example as the broad band between 3700 cm^{-1} and 3200 cm^{-1} (between 2.7 μm and 3.1 μm) which is usually assigned to H-bonded OH stretching vibrations.

Hydrogen discharge lamp (u.v.). Radiation for the ultraviolet region of the spectrum is provided by means of a discharge lamp filled with hydrogen. The upper range of the lamp is about 440 nm and the lower range is 200 nm if the lamp has quartz walls; it can be extended to 180 nm if fused silica walls are fitted. The intensity of the lamp can be almost trebled if deuterium is used in place of hydrogen. *See also* **Tungsten filament lamp.**

Hyperchromic effect (u.v.). If structural modification leads to an increase in the **molar absorption coefficent** for a particular chromophoric group it is said to have brought about an hyperchromic effect. The hyperchromic effect frequently accompanies a **bathochromic shift.** *See also* **Hypochromic effect.**

Hyperfine splitting (e.s.r.). In e.s.r. spectra the hyperfine splitting observed arises owing to interactions between the electron spin and the spins of the adjacent magnetic nuclei. A nucleus possessing a spin quantum number I will split each electronic energy level into $2I + 1$ levels and produce that number of lines in the spectrum. For n equivalent nuclei $2nI + 1$ resonance lines are produced. Both ^{1}H ($I = \frac{1}{2}$) and ^{14}N ($I = 1$) nuclei lead to hyperfine splitting, but ^{12}C ($I = 0$) and ^{16}O ($I = 0$) do not. *See also* **Electron spin resonance** and **Fermi contact interaction.**

Hypochromic effect (u.v.). Structural modification leading to a decrease in the **molar absorption coefficient** for a particular chromophoric group. The hypochromic effect frequently accompanies a **hypsochromic shift**. *See also* **Hyperchromic effect**.

Hypsochromic shift (u.v.). Structural modification in a molecule which leads to a characteristic maximal absorption being displaced towards shorter wavelength (higher frequency) is said to have brought about a hypsochromic shift. *See also* **Bathochromic shift**.

I

I (n.m.r.). *See* **Spin quantum number**.

Incandescence (i.r., pl.). Materials are said to exhibit incandescence if they are self-luminous solely owing to their high temperatures. Such substances produce continuous emission spectra covering a very wide range of wavelengths from the ultraviolet to the infrared. In some cases the distribution of energy in the spectrum approaches that of a black body (*see* **Black body radiation**): in these instances such materials are useful as sources for infrared spectrophotometers.

INDOR (Internuclear double resonance) (n.m.r.). A variation on the simple **multiple resonance** technique specially applicable to weak signals such as ^{13}C satellites. For this purpose the radio-frequency field is adjusted to satisfy the resonance conditions for one nucleus whilst a second radio frequency is used to sweep the spectrum for a second nucleus as the absorption for the first nucleus is observed. If the two nuclei are normally coupled then under these conditions the observed INDOR spectrum, for the first nucleus, consists of a series of negative peaks. This will occur when the second radio-frequency field corresponds to the energy levels of the system being observed.[135,136]

Induction method (Crossed-coil method; Double-coil method) (n.m.r.). A method of detecting n.m.r. absorption signals first used by Bloch.[137] It employs two coils arranged at right angles to each other such that the detector coil picks up signals arising from the re-emission of energy imparted to the nuclei by the irradiation coil (see Figure 25, page 89).

Inductive coupled plasma (ICP) (a.s.). Plasmas are produced by passing radio-frequency discharges through inert gas atmospheres at atmospheric pressures. In the inductive coupled plasma the sample solution is nebulized with nitrogen or argon carrier gas and passed through a quartz tube[138] coaxial with a radio-frequency induction heater supplied with an r.f. signal through a **waveguide**. The r.f. discharge produces a

87

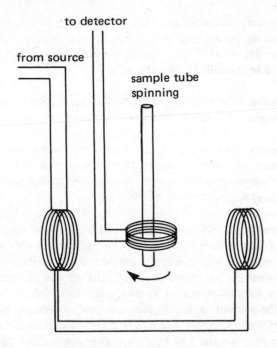

Figure 25. Cross-coil probe

stable plasma with temperatures exceeding 6000 K and gives results which are frequently superior to flame techniques.[139] Detection limits in the parts per billion range have been obtained with precious metals[140] employing a 2 kW argon plasma source. These sources are also being used in emission spectroscopy.[141]

Inelastic mean free path (λ) (Attenuation length; Mean escape depth) (pe.). The depth within a solid from which the ratio $1/e$ of any photoelectrons produced can escape. While the escape depth is equal to λ the normal sampling depth in, for example, **ESCA** is about 3λ.

Infrared detectors. A wide range of infrared detectors exists to cover the wavelengths from near- to far-infrared regions. Recent developments in this field have included detectors operating at low temperatures (superconducting bolometers), and rapid response detectors producing signals faster than 1 ns.[142] *See also* **Bolometer**; **Photoelectric infrared detectors**; **Pneumatic cell**; **Thermocouple**.

Infrared radiation sources. *See* **Globar**; **Nernst glower**; **Nichrome source**; **Opperman source**; **Tungsten filament lamp**.

Infrared spectra. With recent improvements in instrumentation the infrared region of the electromagnetic spectrum (q.v.) is now considered to cover the range from approximately 12 500 to 10 cm^{-1} (0.80–1000 μm), Figure 26, page 90.

It is generally subdivided into three sections: near-infrared, arising from electronic, plus vibrational and rotational transitions, 12,500–4000 cm^{-1} (0.80–2.5 μm), this is also the overtone and combination band (q.v.) region; middle (fundamental) infrared, from vibrational plus rotational transitions, 4000–200 cm^{-1} (2.5–50 μm); far-infrared, rotational transitions, 200–10 cm^{-1} (50–1000 μm).

The middle-infrared region is the one most commonly employed for standard laboratory investigations, the portion between 1428 and 909 cm^{-1} (7.0 and 11.0 μm) being known as the fingerprint region.

Studies beyond 200 cm^{-1} require special instruments and techniques,[143] such as the use of caesium iodide prisms and Fourier-transform procedures (*see* **Fourier-transform spectroscopy**).

Integration (n.m.r.). The process by which the relative areas under spectral peaks are measured and compared. In the case of n.m.r. spectra the integration consists of a series of steps usually run as an overlapping trace to the original spectrum. The heights of the integration steps correspond to the number of nuclei in the groups producing the absorption peaks.

Interfaces (i.r., m.s.). *See* **GC/IR interfaces**; **GC/MS interfaces**; **LC/IR interfaces**; **LC/MS interfaces**.

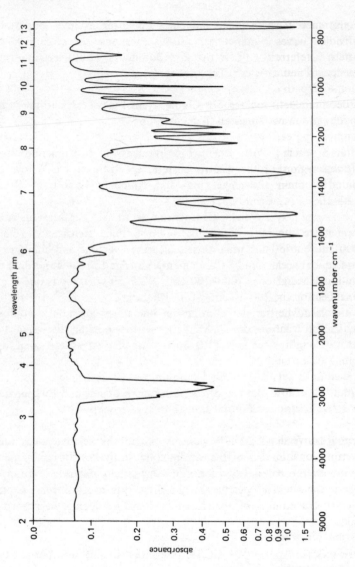

Figure 26. Infrared spectrum of N,N-Dimethylaniline

Interference. If two or more wave motions meet in such a way that the individual waves reinforce or oppose each other the resulting wave pattern is referred to as an interference pattern. If two waves of equal wavelength and amplitude are out of phase by half a wavelength along a common path of motion, they will cancel each other out.

Because electromagnetic radiation is not continuous, but released in quanta, the waves producing interference must be produced from a common source. For this reason a beam splitter is used in interferometers in which this phenomenon is employed.

Fourier-transform infrared **spectroscopy** uses an interference procedure to obtain an interferogram from the spectrum studied. *See also* **Gratings**.

Interference filters. A filter made by sandwiching a thin layer of a transparent dielectric possessing a low refractive index, such as magnesium fluoride, between very thin silver films which are partially transparent. Any incident light that passes through the silver layers undergoes internal reflection between the two silver films, complete interference taking place when the path difference is some multiple of half a wavelength. The transmitted region is a narrow band with its maximum at this wavelength as any other wavelength is reflected away. The maximum transmittance from an interference filter is about 50% with a bandwidth of between 10 nm and 17 nm.

Multi-layer interference filters consist of alternate layers of high and low refractive index materials.

Internal conversion (pl.). The process by which a molecule goes from a low vibrational level of a high energy state to a high vibrational level of a lower energy state with nearly the same total energy. It is from this lower state that a return to the ground state can occur by the emission of **fluorescence**.

Internal reference (n.m.r.). When the n.m.r. reference material (e.g. **TMS** for proton spectra) is dissolved in the same solvent as the sample under investigation it is being employed as an internal reference. Under these conditions it experiences the identical effects of the magnetic field as does the sample. Results are slightly more accurate by using the reference in this way than they are when it is used as an **external reference**.

Internal reflectance spectroscopy (i.r.). *See* **Attenuated total reflectance**.

Internal standard. Quantitative analysis employing internal standards is used in several realms of spectroscopy including infrared, emission and atomic absorption. It involves adding a fixed amount of an internal standard both to a series of increasing concentrations of reference sample and to the unknown concentration. A calibration line is plotted of the ratio of the absorption of the characteristic line of the reference samples and the absorption of the characteristic line of the internal standard against the known concentrations of the reference samples.

This ideally gives a straight-line plot and the unknown sample concentration can be established by determining the position of the corresponding ratio of the sample absorption and internal standard on the calibration line.

In quantitative infrared analysis on solids the internal standard most commonly employed is potassium thiocyanate. It is used as a 1-2% mixture in potassium bromide for making pressed discs (q.v.). The thiocyanate shows a characteristic absorption at 2125 cm^{-1} (4.7 μm) which rarely interferes with absorptions of samples under investigation.[144] The internal standard method has also been applied to **mulls**.[145]

Internuclear double resonance. *See* **INDOR**.

Intersystem crossing (pl.). The population of the **triplet state** can take place by a molecule 'crossing' from a low vibrational energy level of a **singlet state** to an upper vibrational level of a triplet state. Such a process is referred to as an intersystem crossing and takes about 10^{-7} to 10^{-8} s. It eventually leads to **phosphorescence** as electrons can remain for extended periods of time in the triplet state.

Ion. Molecules and individual atoms which have acquired net positive or negative charges through losing or gaining one or more electrons are called ions. A **mass spectrum** is a chart record of ions produced in a **mass spectrometer**. Ions may also be studied by means of **electron spin resonance** spectroscopy.

Ion abundancies (m.s.). Fragmentation of molecules in mass spectrometers gives rise to other positive ions which are recorded as the mass spectra of the molecules. The size of the peak recorded at a particular mass/charge value is related to the number of those particular ions detected at that value. So that the relative heights of the various peaks in any mass spectrum is a measure of the relative ion abundancies. The largest peak in the spectrum is called the **base peak** and all other peaks are expressed as a percentage of the base peak. The relative abundancies of the major peaks in a mass spectrum are represented diagrammatically on a **bar graph**.

Ionization efficiency curve (m.s.). A plot of the relationship between the production of molecule ions (the ionization efficiency) against the electron energy (eV) of the ionization source. Its importance to mass spectrometry is that the ionization efficiency starts to rise at the **ionization energy** of the molecule and reaches a plateau between about 50 to 70 eV. For this reason most mass spectrometers are operated at about 70 eV for general usage as any voltage fluctuation does not affect the reproducibility of the spectra.

Ionization energy (i) (a.s.). The energy necessary to remove an electron from an atom. It is normal to refer particularly to the first ionization energy as that required to remove the most loosely bound electron from the atom to form an ion. In flame photometry ionization of atoms leads to a decrease in the number of neutral metal atoms in the flame and hence results in a corresponding weakness of the atomic spectra lines. At about 2700 °C some elements are almost one hundred per cent ionized.

Ionization energy (ii) (m.s.). The energy required to remove an electron from a molecule varies from one substance to another. For organic compounds the ionization energy is generally between 8 and 15 eV, being the energy needed to create an ion from the molecule existing in the ground vibrational level of the lowest electronic state. This is often referred to as the 'adiabatic ionization energy' as the change from the molecular state to the molecular ion does not usually occur between the lowest vibrational levels; the measured ionization energy will, therefore, differ slightly from this theoretical value.[146]

Ion sources (m.s.). *See* **Electron bombardment ion source; Isatron; Knudsen cell furnace; Laser microprobe; Photoionization source; Spark ionization source; Thermal emission ion source.**

Iris diaphragm. This is frequently used in fluorimeters to control the amount of radiation from the source; the iris diaphragm consists of a series of overlapping plates arranged to move so that a roughly circular opening of variable size is produced. This is achieved by maintaining a nearly constant angle between the edges of adjacent plates.

Irtran (i.r.). The trade name for a series of infrared cell windows made by Eastman Kodak. They are carefully selected fluorides, selenides and sulphides impervious to acids and alkalis for use with aqueous solutions within the spectral range 5000–333 cm^{-1} (0.2–30 μm). They can also be used at elevated temperatures up to 300 °C.

The most general type is made from zinc sulphide with a long-wave transmission limit of 60% for a 1 cm thickness at 645.2 cm^{-1} (15.5 μm).

Isatron (m.s.). A high precision **electron bombardment ion source** used in mass spectrometry.[147] It measures 3 cm long and 2 cm in diameter and is constructed to focus a concentrated electron beam onto the gaseous molecules.

Isobaric fragments (m.s.). On low-resolution mass spectrometers any single peak is likely to represent several different ions formed from the fragmentation process. Ions possessing a common mass/charge ratio, and therefore producing a single peak, are termed isobaric (e.g. N_2^+, CO^+, and $C_2H_4^+$ are isobaric at (m/e) 28).

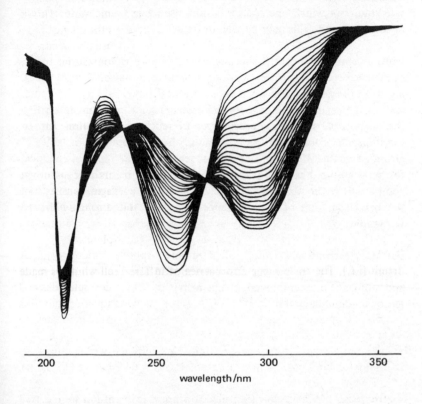

Figure 27. Isobestic points obtained for the rearrangement of a norbornene derivative

Isobestic point. On a series of overlapping spectral curves for a binary mixture obtained with the two components in differing proportions there is one point, and sometimes more, at which all the curves are coincidental. At this isobestic point the absorption is, therefore, a constant for all concentration ratios for the two substances in the binary mixture.

An isobestic point is of particular value as a reference point in quantitative i.r. or u.v. studies and can usually be observed in reaction rate studies in which one species is diminishing and another increasing in concentration. This may be seen in Figure 27.

Isotopic abundance (m.s.). The presence of heavy isotopes leads to the occurrence of mass spectral peaks of higher value than that corresponding to the **molecular ion**. Such peaks are designated $M + 1, M + 2$, etc, and their magnitudes relative to the molecular ion peak are related to the proportion of the higher isotopes present and differ according to the molecular formula of the compound. Isotopic abundance tables[148] giving calculated $M + 1$ and $M + 2$ values for a wide range of chemical formulas are used as a guide to the true chemical formula of unknown compounds by comparison with the sizes of the peaks measured from the mass spectrum of the substance. *See also* **Abundance tables** and **Bar graph**.

IUPAC (International Union of Pure and Applied Chemistry). An organization responsible for coordinating the standardization of signs and nomenclature employed in chemistry in order to improve international communication.[149,150]

J

J (n.m.r.). The symbol used to denote the **spin–spin coupling** constant and represents the magnitude of the separation between the peaks of a multiplet arising from the coupling. The value of *J* is always expressed in **hertz**.

Jet separator (m.s.). One of the earliest **GC/MS interfaces** was the two stage Becker-Ryhage jet[151] in which enrichment occurs due to loss of carrier gas as the eluate passes along small jets separated by short distances from each other. About 50 per cent of the sample passes to the mass spectrometer and the carrier gas content is reduced to 2 per cent. *See also* **Watson–Biemann separator**.

K

K_{α}. *See* **X-rays.**

Karplus equation (n.m.r.). The dihedral angle (Φ) between vicinal protons can be calculated approximately from the coupling constant of the two protons by the use of this equation originally proposed by Karplus.[152,153]

$$J = 8.5 \cos^2 \Phi - 0.28 \quad \text{if } 0° \leqslant \Phi \leqslant 90°$$
$$J = 9.5 \cos^2 \Phi - 0.28 \quad \text{if } 90° < \Phi \leqslant 180°$$

The calculation must be modified to allow for the effects of other substituents close to the vicinal protons and several variations on the equations have been developed.

Kayser K (i.r.). A term no longer recommended for spectroscopic measurements. It was originally suggested for the unit of **wavenumber** by the Joint Commission for Spectroscopy but considerable controversy arose,[154,155] and it has not been adopted as an **SI unit** although it is found in older literature.

$$1 \text{ K} = 1 \text{ cm}^{-1}$$

K bands (i) (pe.). *See* **X-rays.**

K bands (ii) (u.v.). A system of designation[156] of bands in ultraviolet/ visible spectroscopy uses this symbol to refer to absorptions arising from $\pi \rightarrow \pi^*$ transitions usually having high molar absorption coefficients, $\epsilon > 10\,000$. These occur on conjugated systems, particularly aromatic compounds. (K is from the German 'kongugierte'. *See also* **Bands (ultraviolet).**

KBr pellet (i.r.). One method of studying solid samples in infrared spectroscopy is by preparing a **pressed disc,** by compressing a finely ground mixture of the sample and potassium bromide. As the potassium bromide does not absorb in the middle-infrared region the only absorption obtained is that due to the sample. *See also* **Mulls.**

Klystron oscillator (e.s.r.). Signal generators employed in electron spin resonance spectrometers are usually of the klystron type operating in the Q (8 mm) of X (3 cm) microwave bands. The klystron is a vacuum tube which produces an electron beam from a heated cathode. A high positive potential attracts this beam into the microwave cavity in which electric field variations create oscillations which lead to bunching of the electrons in place of the formerly homogeneous beam. From the resonance cavity the electron bunches pass to a repeller electrode, negative with respect to the cathode, which returns them to the cavity.

Oscillation of the system is maintained by employing suitable voltages and energy is then fed from the klystron to a **waveguide** which conducts it to the **hybrid T** from which it passes to the sample under investigation in the magnetic field.

The klystron is also used in conjunction with the electrodeless discharge tube source in atomic absorption spectroscopy.

Knudsen cell furnace (m.s.). A device used to produce a fine molecular beam for study by mass spectrometry. It consists of a small tungsten or tantalum crucible contained in a high-temperature furnace surrounded by a water-cooled housing. The sample is heated to 2500 °C by electrical heating under high vacuum and the resulting stream of molecules is directed through a minute effusion hole (< 1 mm in diameter) across the line of direction of the electron beam in the ionization chamber.[157]

It is used particularly for thermodynamic studies on solids at elevated temperatures, and has been employed for vapour pressure investigations on silver and graphite.[158]

Koopman's rule (pe.). Interpretation of photoelectron spectra is sometimes made in terms of orbital electronic structures of the neutral system.[159] This is based upon the assumption that the ionization energy, I_k, of a species, k, is identified by the relationship:

$$I_k = -\epsilon_k^{scf}$$

where ϵ_k^{scf} is the self-consistent field orbital energy.

It applies only to closed-shell systems and in general gives high values for ionization energies because it neglects the effect of orbital relaxation upon ionization. As a result the calculated ionization energies may be more than 10 per cent in error.

KRS-5 (i.r.). Specially developed for use as a prism medium for infrared spectrometers KRS-5 is a mixture of thallium bromide and thallium iodide with molar percentages of 45.7 and 54.3 respectively.[160] *See also* **Prisms, infrared.**

L

L_α. *See* **X-rays.**

Lambert law (i.r., u.v.). For homogeneous substances the portion of incident radiant energy absorbed is proportional to the thickness of the absorbing material.[161,162]

Mathematically this is expressed as

$$-\frac{\mathrm{d}I}{\mathrm{d}l} \propto I$$

which on integration becomes

$$\log\frac{I_0}{I} = k_1 l$$

See also **Beer–Lambert laws** and **Absorbance.**

Landé g factor (Spectroscopic splitting factor) (e.s.r., n.m.r.). A dimensionless quantity characteristic of the particle being studied and calculated from the equation

$$g = 1 + \frac{J(J+1) + S(S+1) - L(L+1)}{2J(J+1)}$$

It relates the magnetic moment of the particle and the angular momentum according to the relationship

$$\mu = \frac{geh}{4\pi m_p c}\{J(J+1)\}^{1/2} = g\mu_N\{J(J+1)\}^{1/2}$$

$g \approx 2$ for particles with pure electronic spin.
$g \approx 1$ for particles with pure electronic orbital magnetism.
More exactly $g_e = 2.002\,319$ for unbound electrons and variations of

101

the *g* factor from this value are often considered as being analogous to chemical shifts in n.m.r. as they depend upon the chemical environment of the unpaired electron. For most nuclei g_N has values between -2 and $+6$.

Larmor frequency (ω_L) (n.m.r.). A name for the angular frequency of electron **precession**. It is calculated from the following equation

$$\omega_L = \frac{e\mathbf{B}}{2m_e}$$

It is now usual to differentiate between the Larmor frequency — applicable to electrons — and the **nuclear angular precession frequency** for nuclei.

Laser (r.s.). An acronym constructed from the expression Light Amplification by Stimulated Emission of Radiation which roughly describes the nature of the devices.[163] The laser may take many forms, but the simplest is the ruby laser consisting to all intents and purposes of an aluminium oxide matrix containing some Cr^{3+} ions. The crystal is highly polished at each end and acts as an optical resonator.

Energy from a flash lamp passes into the laser rod or tube; this leads to a rapid build-up of energy and the creation of metastable states eventually producing a **population inversion**. The reversion of one active species automatically stimulates other active species to release energy and a rapid amplification occurs. The powerful beam of radiation escapes as a burst of energy from one end of the rod. The resulting beam is monochromatic, coherent and parallel (*see* Figure 28).

Many lasers are now made using gases and dyes rather than crystals. These have the advantage that they can produce a number of different wavelengths; the argon laser, for example, has 8 different radiations. Gas lasers are often preferred for their continuous steady power[164] (the helium-neon laser, for instance, gives less than 1 per cent variation in output). Lasers have also been produced from organic liquids such as benzene and toluene, whilst semiconductor lasers made from gallium arsenide and similar materials are able to convert electricity directly

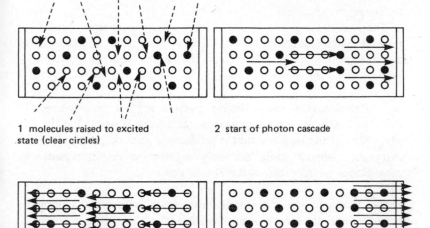

1 molecules raised to excited
state (clear circles)

2 start of photon cascade

3 photons reflected back and
forth

4 beam passes out through the
semi-silvered end

Figure 28. Operation of a laser

to coherent light beams. Lasers based upon dyes, such as Rhodamine
6G are tuneable over a range of 60–70 nm in the visible and near
infrared regions.[165]

Laser beam excitation. A general expression employed when lasers are
used as a means of heating small samples in order to bring about excitation or ionization.

Laser microprobe (em., m.s.). Sample excitation for emission spectroscopy applicable to non-conducting materials can be achieved by
focusing the beam from a **laser** onto a small amount of sample. About

20% of the laser beam energy is used in exciting the atoms in the material studied;[166] the vapour from the sample is used to short circuit the gap between the electrodes on the emission spectrometer thus producing the spectrum.

The laser microprobe is also being used in mass spectrometry to vaporize sample directly into the ion source.[167,168] A spot size of up to 800 μm is made on a pellet of the sample held on the end of a probe inserted through a vacuum lock.

Laser Raman spectroscopy. Raman spectroscopy has advanced rapidly since the laser was first introduced as the source of excitation.[169] The advantages of the laser are that it produces a powerful beam of monochromatic, coherent radiation vastly superior to that obtained with other Raman sources such as the **Toronto arc**.

Lattice (n.m.r.). The total environment around a nucleus including the adjacent molecular structure and the solvent.

L bands. *See* **X-rays.**

LC/IR interfaces. The main problem associated with interfacing liquid chromatography with infrared spectrometers has been in obtaining a rapid enough scan of the spectrum as substances are eluted from the column and through the detector cell. The development of Fourier-transform techniques (*see* **Fourier-transform spectroscopy**) has greatly assisted in overcoming this difficulty[170] so that the continuous monitoring of eluates is now possible using flow through cells of very small capacity.

LC/MS interfaces. The main problem in coupling liquid chromatography with a mass spectrometer is in the removal of the large quantity of solvent.[171] In one approach to the problem[172] the sample with the major part of the solvent is led into the spectrometer source via a capillary inlet. In this application the solvent serves as a **chemical ionization** reactant gas so that proton transfer assists ionization of the solute molecules.

An alternative treatment[173] has been to pass all the eluate on a continuously moving polymeric belt forming part of a **direct inlet system**. The solvent is rapidly removed by a strong infrared lamp and the sample flash evaporated within the ion source chamber. A recent technique[174] has been to concentrate the sample by allowing it to flow down an electrically heated wire and through a regulatory needle valve from which it is sprayed into the ion source. This is claimed to provide a twenty-fold increase in concentration.

Lead sulphide conductive cells. *See* **Photoelectric infrared detectors.**

Leak valve (m.s.). *See* **Molecular leak.**

LEED. *See* **Low energy electron diffraction.**

Ligand field theory (u.v.). When a transition metal ion is attached to water molecules, or any other ligand, electronic transitions can occur between the different groups of degenerate orbitals. This absorption of energy gives rise to the characteristic colours of transition metal compounds that are studied in ultraviolet/visible spectroscopy. This explanation of the colour of these compounds is the basis of the ligand field theory.[175]

Light. The specific part of the **electromagnetic spectrum** covering the visible region that is normally associated with human vision. It extends from the violet (380 nm) to the red (780 nm) and is commonly studied jointly with the ultraviolet region (*see* **Ultraviolet/visible spectrum**). *See also* **Velocity of light.**

Light (monochromatic). Light of one wavelength, which is the ideal state required after monochromation has been carried out in the spectrometer. In most cases this is rarely attained, as irrespective of the quality of the optics in the instrument there is still a finite slit width. As a result most so-called monochromatic light actually covers a narrow band on either side of the required line. *See also* **Effective bandwidth.**

Line spectra (a.s., em.). The characteristic spectra of atoms resulting from charges in the energy levels of the electrons. They consist of a large number of irregularly spaced lines. *See also* **Atomic spectra**.

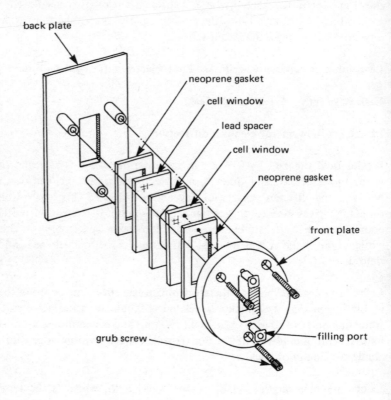

Figure 29. Fixed path-length infrared cell

Liquid cell (i.r.). Liquid cells for infrared spectroscopy can be of fixed or variable path length. A substance of molecular weight about 200 in 1% solution gives satisfactory spectra with 1.0 mm path length; a

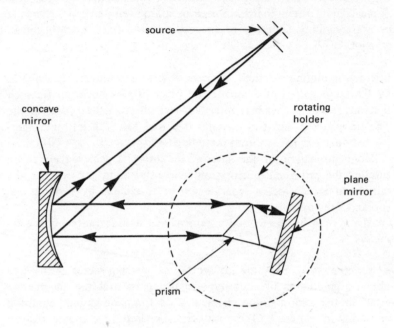

Figure 30. Littrow mounting

standard cell of this thickness is filled by 0.4 mL of solution. For experimental reasons cells can be either permanently built or demountable for assembly when required. A variety of spacers is available permitting selection of path lengths between 0.05 mm and 3.0 mm. Cells are normally designed for syringe filling. Variable path length cells, for accurate compensation work and for quantitative studies, are designed around a vernier screw thread which enables the path length to be adjusted at will. *See also* **Microcells**.

Liquid filters (r.s.). Monochromation of radiation from arc and lamp sources in Raman spectroscopy is carried out using various solutions to absorb unwanted radiation and to isolate the required exciting line. Typical liquid filters that have been used are saturated sodium nitrate or praseodymium chloride in aqueous solution and p-nitrotouene in ethanol.

Littrow mounting. Double dispersion and resolution are achieved by the combination of a mirror and prism as a monochromator such that any refracted beam is reflected back through the prism. In the case of quartz prisms this obviates the need for both left- and right-handed quartz as in the **Cornu mounting** (see Figure 30, page 107).

Monochromation in the Littrow mounting is achieved either by rotating the prism and mirror simultaneously, or by just rotating the mirror. In some cases a separate mirror is avoided by silvering one surface of the prism.

The system has been used extensively in a wide range of spectrometers.

Log sector (em.). Both the log sector and the **step sector** are used to obtain a graduation of intensity of emission spectral lines in order to establish the exposure/line characteristics for photographic emulsion on emission plates.[176] They can also be used to compare relative intensities of different lines and for quantitative emission spectroscopy (*see* Figure 31).

Lone pair electrons. Electron pairs which, while not involved in bond formation in a particular compound, are capable of taking part in forming a coordinate bond with a substance possessing an incomplete octet. Ammonia is a typical example of a substance possessing a lone pair and is capable of forming a bond with compounds such as boron trifluoride.

$$
\begin{array}{cc}
\text{H} & \text{F} \\
\text{x·} & \text{x·} \\
\text{H} \overset{\text{·}}{\underset{\text{·}}{\text{x}}} \text{N:} + \text{B} \overset{\text{·}}{\underset{\text{·}}{\text{x}}} \text{F} \longrightarrow \text{H}_3\text{N: BF}_3 \\
\text{·x} & \text{·x} \\
\text{H} & \text{F}
\end{array}
$$

slit of the spectrograph

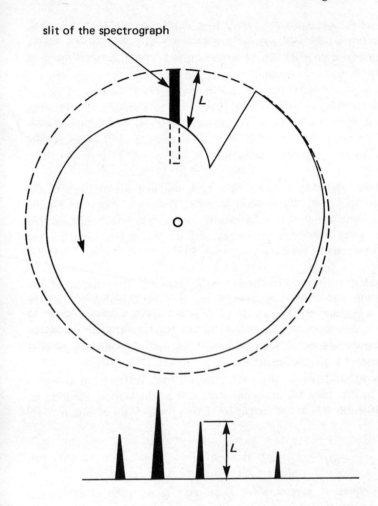

Figure 31. Log sector (appearance of spectrum line of different
intensity indicated below)

Longitudinal relaxation (T_1) (Spin-lattice relaxation (n.m.r.). A process by which nuclear spin energy is lost and occurs by transfer to the molecular environment. The energy is lost by localized variable magnetic fields to other magnetic nuclei. Impurities in the n.m.r. sample, particularly paramagnetic species, greatly increase the rate of relaxation. Broadening of absorption peaks due to spin–lattice relaxation occurs if T_1 is less than 10^{-3} s. Liquid samples rarely exhibit broadening of the absorption peaks as T_1 in these cases is usually greater than 10^{-2} s. *See also* **Transverse relaxation.**

Long-range coupling (n.m.r.). **Spin–spin coupling** for protons usually extends only over three bond lengths. However, coupling beyond this is possible and occurs frequently in unsaturated systems, such as allenes. Over this extended range, of four or five bond lengths, the coupling constants are usually less than 3 Hz.

Lorentzian curve (distribution) (e.s.r., n.m.r.). The shapes of the absorption and dispersion lines which may be calculated from the **Bloch equations** are known as Lorentzian curves and are those expected under ideal circumstances. In practice the theoretical shapes are rarely obtained in n.m.r. spectra due to instrumental variations and sample inhomogeneities.

Lorentzian (also known as Cauchy) curves differ from **Gaussian curves** in that they tail more on either side of the central distribution, the half-width W/2 at half-height $H/2$ being given by the equation

$$\frac{W}{2} = \frac{1}{\pi H}$$

The equation governing the Lorentzian distribution of the probability at a value x for a set of variables is

$$P(x) = \frac{1}{\pi} \frac{a}{a^2 + (x - b)^2}$$

where $a > 0$, and $-\infty < x < +\infty$.

Low energy electron diffraction (LEED) (pe.). A term applied to photoelectron spectroscopy involving electrons of low power. It is used particularly for the study of surface layers.[177] Electrons with energies between 0–500 eV are diffracted from the crystal surface. Only those electrons that have lost less than 2 per cent of their power are allowed to pass to the detector and this accounts for about 1 per cent of the electrons focused from the **electron gun**. Data is usually presented in the form of intensity/energy curves with the electron beam energy being scanned over a 300 eV range.[178] LEED has been responsible for showing substantial differences between surface structures and those occurring at depth in materials.

Luminescence (pl.). A collective word used to cover all forms of light emission other than those arising from elevated temperature. Any system exhibiting luminescence is losing energy in the form of emitted radiation. It is subdivided into **fluorescence** in which the excited state exists for less than 10^{-8} s, and **phosphorescence** in which the excited state lasts for more than 10^{-8} s. All processes of luminescence involve three steps: absorption, change in energy state and emission, which may be brought about by a wide variety of processes (e.g. **photoluminescence** by photons and **chemiluminescence** in chemical reactions).

Lyman series (em.). *See* **Atomic spectra**.

M

McLafferty rearrangement (m.s.). Fragmentation in mass spectrometry frequently involves rearrangement processes within the ions created. The McLafferty rearrangement[179] is a six-centre transfer of a gamma hydrogen to a double bonded atom followed by beta-bond cleavage.[180] This is shown by the following process with methyl propyl ketone (pentan-2-one)[181]

Magic T (e.s.r.). *See* **Hybrid T.**

Magnetic deflection mass spectrometer (m.s.). Mass spectra can be obtained by the exclusive use of a magnetic field for resolution of the accelerated positive ions as in a **single-focusing mass spectrometer**, or by using the magnetic field in conjunction with an electrostatic field in a **double-focusing mass spectrometer**. The angles of deflection used for the magnetic fields may be $60°$, $90°$, $120°$ or $180°$, the ions being separated according to their mass/charge ratios satisfying the equation

$$\frac{m}{e} = \frac{H^2 r^2}{2V}$$

Individual ionic species are brought to a focus by varying either **H** or **V**.

Magnetic field strength (H). The intensity of a magnetic field at a point and the force that would exist on a unit north magnetic pole at that point. It is related to the **magnetic flux density** and permeability by the equation

$$H = \frac{B}{\mu}$$

In **SI units** the magnetic field strength is measured in ampere per metre. 1 A m^{-1} corresponding to 4×10^{-3} oersted.

Magnetic flux (Φ). *See* **Weber.**

Magnetic flux density (Magnetic induction). **(B).** The magnetic flux per unit area perpendicular to the lines of flux. The **SI unit** for this quantity is the tesla, defined as kg s^{-2} A^{-1} = Wb m^{-2} = V s m^{-2}

$$1 \text{ T} = 10^4 \text{ gauss}$$

Magnetic induction. *See* **Magnetic flux density.**

Magnetic moment (μ) (n.m.r.). The turning force (torque) operating on a magnet when it is held at right angles to a field of unit intensity and is equal to the product of the magnetic pole strength and the distance between the poles. It is measured in ampere per square metre.

$$1 \text{ A m}^{-2} = 10^3 \text{ e.m.u. s}^{-1}$$

Magnetization (M). The magnetic moment of a magnet per unit volume, given by the equation

$$M = \frac{B}{\mu_0} - H$$

Magnetogyric ratio (γ) (n.m.r.). The ratio between the **magnetic moment** and the angular momentum:

$$\gamma = \frac{\mu 2\pi}{Ih}$$

Under the SI scheme (*see* **SI units**) the magnetogyric ratio is measured in kg s^{-3} A^{-1} = s^{-1} T^{-1}. For protons the value of γ_p is 2.6752 × 10^8 s^{-1} T^{-1}.

Magneton. *See* **Bohr magneton** and **Nuclear magneton.**

Marker peaks (i.r.). *See* **Calibration peaks.**

Maser (r.s.). An acronym for Microwave Amplification by Stimulated Emission of Radiation, originally coined with reference to the precursors of the modern **lasers.**

Mass analyser (m.s.). The portion of the mass spectrometer that produces the resolution of the positive ions. In many instruments this will be simply a magnetic field, although in a **double-focusing mass spectrometer** it will also include an electrostatic field. The mass analyser in quadrupole and **time-of-flight mass spectrometer** is a horizontal tube known as the drift tube (*see* **Drift region**).

Mass/charge ratio (m.s.). A **mass spectrometer** actually measures mass/charge ratios rather than absolute mass values and it is these that are recorded on the mass spectrum. As the majority of positive ions produced carry a single charge there is usually no difficulty in interpreting the spectra. However, this is not invariably the case and most spectra of organic compounds include some peaks due to doubly (or even triply) charged ions. In these cases the spectral peak occurs at a half (or one-third) the true mass value. The existence of a double charged ion can easily be determined in those cases in which the true mass value is an odd number as the observed mass/charge ratio is found at a half mass charge. *See also* **Metastable ions.**

Massmann furnace (a.s.). A flameless method of atomization used in atomic absorption spectroscopy which is a carbon tube atom reservoir and was the forerunner of the present heated graphite atomizers. It consists of a hollow graphite cylinder about 50 mm long and 10 mm in diameter situated so that the radiation beam passes along the axis of the tube.[182] The sample (1–200 μL) is placed in the middle of the tube through a small hole. By passing a small current through the graphite tube the sample is first evaporated, then atomized by applying a much higher current (500 A at 10 V). Oxidation of the graphite is prevented by maintaining it in a nitrogen or argon atmosphere. *See also* **Carbon filament atom reservoir**; **Delves cup**; **Tantalum boat**.

Mass spectrograph (m.s.). An instrument in which positive ions are separated according to their mass/charge ratios and used to produce images on a photographic plate. Such instruments are commonly used to study isotopic abundancies in elements. The early mass studies carried out by Thomson[183] and Aston[184] used photographic plates to detect the ions.

Mass spectrometer (m.s.). In a mass spectrometer the positive ions are separated according to their mass/charge ratios and the individual ionic species detected and recorded electronically.[185] The various ions are indicated on the spectral chart as peaks of different sizes corresponding to the quantities of the individual ions produced. The spectrometer consists of the following distinct parts: the inlet system, the **ion source**, the **accelerating slits**, the **mass analyser**, the ion collector and the recorder.

Mass spectrum (m.s.). This may be either in the form of a photographic record, as produced by a **mass spectrograph**, or as a chart record of the type obtained from a **mass spectrometer**. In either case the spectrum is a record of the mass/charge ratios of the positively charged ions plotted against their relative abundancies. As the size and nature of the ions produced is dependent upon the molecular structure the mass spectrum is characteristic for that substance under the defined spectrometric conditions. *See also* **Bar graph**.

Mattauch–Herzog mass analyser (m.s.). An arrangement for a **double-focusing mass spectrometer** which consists of a $31^\circ\ 50'$ electrostatic field deflecting the ions in one direction followed by a 90° magnetic field turning them back in the opposite direction. The electrostatic field serves to produce a beam of ions possessing uniform energies which are then focused according to their mass/charge ratios by the magnetic field.[186] *See also* **Nier–Johnson mass analyser**.

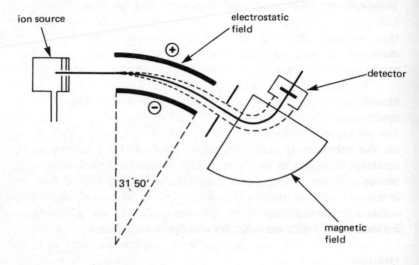

Figure 32. Mattauch–Herzog geometry for double focusing mass spectrometer

Maxwell–Boltzmann distribution and statistics. Maxwell–Boltzmann statistics is the name given to the distribution equations developed separately by Maxwell and Boltzmann before the introduction of quantum mechanics. Several different forms of the equations exist,[187] the most common being that the number of molecules $N(v)$ with velocities between v and $v + dv$ is given by

$$N(v) = 4\pi N \left(\frac{m}{2\pi kT}\right)^{3/2} e^{-mv^2/2kT} v^2 \, dv$$

See also **Boltzmann principle** and **Boltzmann ratio**.

Mean escape depth (pe.). *See* **Inelastic mean free path**.

Meker burner (a.s.). *See* **Burner**.

Membrane separator (m.s.). One form of GC/MS interface is the use of thin silicone rubber membranes which permit the preferential penetration of organic solutes. Up to 90 per cent of the substance passes to the mass spectrometer with a substantial reduction in the proportion of carrier gas. *See also* **Jet separator** and **Watson–Biemann separator**.

Mercury diffusion pump (m.s.). Maintenance of high vacua in mass spectrometers is achieved by using a combination of an oil pump (the fore-pump) and a mercury diffusion pump. The mercury pump works on the principle of heated mercury vapour being condensed by a cooling system and in so doing dragging with it the molecules of gas remaining in the system. For this purpose the mercury diffusion pump is situated between the backing pump and the evacuated vessel. The condensed mercury runs towards the backing pump and is eventually led back to the boiler reservoir. *See also* **Sputter-ion pump**.

Metastable ions (m.s.). Mass spectra (*see* **Mass spectrum**) frequently include diffuse peaks possessiong non-integral mass/charge values.[188] These arise from ions which have decomposed after they have been resolved by the analyser. As such they have been accelerated as one species (the metastable ion) and resolved as another (the daughter ion). The process is

$$A^+ \longrightarrow B^+ + C$$

accelerated   detected   neutral

The apparent mass (M^*) observed on the spectrum is related to the masses of A^+ and B^+ by the relationship

$$M^* \approx \frac{M_B{}^2}{M_A}$$

The precursor ion (A^+) is sometimes known as the parent ion — this should not be confused with the **molecular ion** which produces what has previously been called the parent peak on the mass spectrum.

The calculation of the mass/charge values for the precursor ion and the detected ion from the above formula has been greatly facilitated by publication of a list of metastable transitions.[189] The study of metastable ions has become a specialization in itself and **chemical ionization** techniques coupled with long **field-free regions** in double-focusing mass spectrometers have enabled a high sensitivity for investigation to be achieved.[190]

Microcells (i.r.). For micro-scale infrared spectroscopy special microcells are prepared by drilling a small cavity in a solid block of sodium chloride or potassium bromide.[191] By this means it is possible to prepare a cell with a path length of < 5 mm possessing an internal volume of about $0.2\,\mu L$. Silver chloride has also been used for this purpose.

An alternative type of microcell[192] uses small silver chloride windows at the end of a silver or Teflon tube with a capacity of $30\,\mu L$.

Micrometre (μm). The recommended **SI unit** to take the place of the micron (μ).

$$1\,\mu m = 1 \times 10^{-6}\ m = 1 \times 10^{-4}\ cm$$

Micron. *See* **Micrometre.**

Microprobe (em.). *See* **Laser microprobe.**

Microwave powered sources (a.s.). *See* **Electrodeless discharge tubes.**

Microwaves. The microwave region of the **electromagnetic spectrum** covers the wavelength range between 1 mm and 30 m. This apparent misnomer, as regards spectroscopy, is because it is the micro (or short)-wave region for the purpose of radio engineers.

Microwave spectroscopy. Transitions between different rotational energy levels of the same vibrational and electronic states are studied by the use of microwaves. Simple symmetric molecules such as CCl_4 do not give microwave spectra as it requires the existence of a permanent dipole moment in the molecule.

The microwave region employed is that between 1 mm and 30 cm (1000 MHz–300 GHz) and necessitates the use of klystron oscillators and **waveguides**. The absorption cell for these studies is a piece of waveguide ten feet long sealed at each end with mica windows.[193]

Millimicron. *See* **Nanometre**.

Modulation. The general application of the word is to a wave motion which has a second wave motion superimposed upon it, and is used particularly in radio transmission in which audiofrequency waves are transmitted by combining their wave motion with that of a continuous wave of constant amplitude and frequency.

In spectroscopy, modulation is brought about by either mechanical or electronic processes which cut the electromagnetic radiation beam at a constant rate with frequencies of up to 1000 Hz. The main advantages of this are that an a.c. amplifier can be used that is tuned to the modulated signal frequency such that stray signals are blocked giving an improved spectrum.

Molar absorption coefficient (ϵ) (Molar extinction coefficient; Molar absorbtivity) (i.r., u.v.). The recommended term for the **absorbance** for a molar concentration of a substance with a path length of 1 cm determined at a specified wavelength. Its value is obtained from the equation

$$\epsilon = \frac{A}{cl}$$

Strictly speaking, in compliance with SI units the path length should be specified in metres but it is current general practice for centimetres to be used for this purpose.

Under defined conditions of solvent, pH and temperature the molar absorption coefficient for a particular compound is a constant at the specified wavelength. *See also* **Beer–Lambert law.**

Molecular ion (m.s.). Removal of an electron from a molecule by electron bombardment in the ion source produces the molecular ion of the species; this is detected at a mass/charge ratio that corresponds to the molecular weight of the molecule. Peaks obtained from molecular ions are designated by the letter M on the **bar graph** or mass spectrum; in earlier literature the term parent peak and letter P have been used for this.

Molecular leak (m.s.). To prevent the vacuum in a mass spectrometer from being destroyed by too rapid a flow of sample and to ensure that the sample in the ionization chamber is representative of the sample in the reservoir it is slowly bled from the reservoir to the ionization chamber through a small adjustable orifice known as a leak. By this means the ionization chamber is kept at a pressure of 10^{-5} torr or less. A wide variety of leaks have been used including porcelain rods, glass diaphragms, quartz tubes and capillary tubes.

Molecular spectra (i.r., u.v.). An inclusive expression referring to spectra arising as a result of transitions in molecules. This includes **electronic spectra**, studied in the ultraviolet/visible region, and vibrational or rotational spectra studied in the infrared region (*see* **Infrared spectra**).

Monochromatic radiation. Originally applied to visible light, it is now applied to any electromagnetic radiation in which the waves all have the same wavelength. The literal meaning of the term is 'one colour'.

Monochromator. An instrument used to select **electromagnetic radiation** of a single frequency from a beam of radiation covering a wide range. This is achieved by employing the dispersion properties of either a prism or a **grating**. Some instruments may use both of these in combination, as in the **double monochromator** or separately in special mountings such as the **Littrow mounting.**

Monochromation of **X-rays** for **photoelectron spectroscopy** is most commonly achieved by using the principle of diffraction through quartz crystals based upon the **Bragg equation**. As diffraction of the X-ray beam will only occur for those wavelengths and angles which satisfy the equation it is possible to achieve a reflectivity of about 45 per cent for the selected X-rays.[194]

Mössbauer spectroscopy. Mössbauer[195] found that in solid crystalline media the absorption or emission of γ radiation by nuclei occurred without the loss of recoil energy. This enabled nuclear energy levels to be studied. Conditions employed for the study require the γ-ray source to move relative to the absorbing species so that the **Doppler effect** produced adds energy to the system.[196]

Mössbauer (or γ-resonance nuclear fluorescence) spectroscopy has many applications in qualitative and structural analysis[197] and the effect has been observed with 42 elements.[198] Its main use, however, is in the study of iron and tin compounds where it is used for the measurement of nuclear hyperfine structure as well as for quantitative analysis.[199]

Mulls (i.r.). One method of obtaining infrared spectra from solid materials is to grind the powdered material with a liquid specially chosen to absorb only at unimportant sections of the spectrum. A fine dispersion of the solid in the mulling agent is produced which is then squeezed into a thin film between potassium bromide plates and placed in the infrared beam of the spectrometer. The most common mulling agent is liquid paraffin sold under the trade name **Nujol**; and to obtain the spectrum in the region covered by the Nujol absorptions it is necessary to run a second mull prepared in a different liquid possessing no C−H bonds. Hexachlorobutadiene or **Fluorolube** are used for this purpose.

Only small quantities of material are required to make a mull, about 3 to 4 mg of the solid being ground with two or three drops of the mulling agent in an agate pestle and mortar.

Multi-centre fragmentation (m.s.). Electron bombardment of a molecule in the ion source may lead to cleavage of several bonds, simultaneously

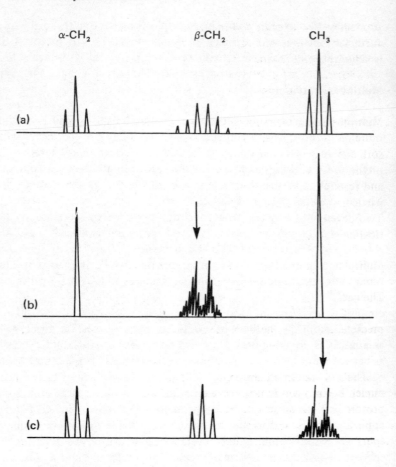

Figure 33. Multiple resonance. Spectra of n-propyl bromide,
$CH_3CH_2CH_2Br$: (a) normal spectrum, showing spin–spin splitting;
(b) effect of irradiating at frequency of the β-CH_2 group: α-CH_2
and CH_3 are collapsed to single lines; (c) effect of irradiating at
frequency of CH_3 group: β-CH_2 becomes a triplet, α-CH_2 is
unaffected. In each case the large arrow indicates the point of
irradiation, and the spectra are frequency-swept
ones

producing a number of fragmentation ions, each of which can undergo further elimination and rearrangement process. Such a result is known as a multi-centre fragmentation.

Multiple quantum transitions (n.m.r.). *See* **Combination lines.**

Multiple resonance (n.m.r.). The application of two (double resonance) or more radio frequencies adjusted to enable several groups of nuclei to absorb energy at the same magnetic field strength. This leads to the collapse of the normal **spin–spin coupling** multiplicities for the irradiated nuclei and results in more simple spectra[200] (see Figure 31, page 122). It is of particular value in structural determination of complicated molecules.[201] Two developments of the multiple resonance technique are **INDOR** and **tickling.**

Multiplets (n.m.r.). **Spin–spin coupling** between nuclei leads to simple n.m.r. absorption peaks being split into a number of smaller peaks. The nature of the multiplets produced is an indication of the number of coupling nuclei, and in the case of first-order spectra of protons can be predicted from the **Pascal triangle.** The formation of multiplets occurs as a result of coupling between nuclei possessing different spins as well as between nuclei possessing identical spins.

The multiplicity of any absorption is established by the neighbouring nuclei. For protons it has been determined that n magnetically equivalent protons will give rise to a multiplicity of $n + 1$. The general equation applicable to any coupling nuclei is that n equivalent nuclei possessing a spin I will produce a multiplicity of $2nI + 1$. *See also* **Coupling constant.**

Multiplicity (pl.). *See* **Triplet state.**

Multiply charged ions (m.s.). Ions produced in a mass spectrometer may carry more than one positive charge; these multiply charged ions are detected as a fraction of their true mass value as the spectrometer is actually measuring the mass/charge ratio. Thus doubly charged ions occur at half the true mass value and triply charged ions at one-third the true values. Such ions arise most readily with unsaturated organic compounds.

N

Nanometre (nm). Under the scheme of **SI units** the nanometre has superseded the former expression millimicron (mμ). This dimension is used particularly for the ultraviolet/visible region in the range below 800 nm.

$$1 \text{ nm} = 10^{-9} \text{ m} = 10\text{Å}$$

Near-infrared region. *See* **Infrared spectra.**

Nebulizers (a.s.). Production of aerosols for atomic absorption spectroscopy is usually carried out by pneumatic devices called nebulizers (the word atomizer should not be used in this sense).[202] The sample to be nebulized is drawn up a capillary either by the action of a current of air across the top of the capillary or by the passage of oxygen through a capillary concentric with the sample capillary (see Figure 34, page 125). A high-pressure gas flow is necessary to produce the fine aerosol that is fed into the flame. Droplet sizes range from 1 to 50 μm and only about 10 per cent of the sample actually reaches the flame.[203] For use with acidic solutions nebulizer capillaries are made from platinum-iridium alloys, whilst the anulus is formed from tantalum.

Negatively charged ions (m.s.). Although mass spectrometers are normally used to examine positive ions it is possible to investigate the nature of any negatively charged ions formed by the ionization process by reversing the electric and magnetic fields in the source and the mass analyser. Negative ions occur most readily with elements possessing high electron affinities as the negative ions are frequently formed by electron capture processes. Spectra obtained from negative ions are weaker than are those from positive ions.

Nernst glower (i.r.). One of the four type of source of infrared radiation used in spectrometers. It consists of a rod or hollow tube made from zirconium and thorium oxides[204] that glows when heated electrically to

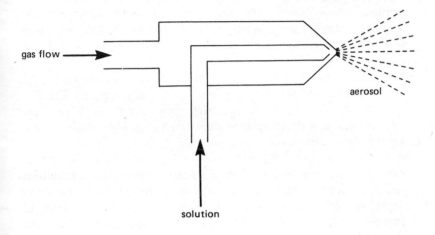

gas flow →

aerosol

solution

Figure 34. Nebulizer

about 1800 °C. Other rare earth oxides may also be used in its formation. Because it is virtually a nonconductor at low temperatures it has to be preheated to 800 °C before further raising of the temperature can occur from the passage of electricity through the rod.

For most purposes the Nernst glower is about 3 cm long and 2 to 3 mm in diameter with platinum electrodes attached to each end. *See also* **Globar**; **Nichrome source**; **Opperman source**.

Neutral fragments (m.s.). In the formation of positive ions in mass spectrometry other ions and molecular fragments are also formed. Some of these carry no charge at all and cannot be resolved by the spectrometer. The nature of the neutral fragments can only be inferred from the positive ions that are formed and from the existence of **metstable ions**.

Neutron. Of the three fundamental particles which go to make atoms the neutron is the only one which carries no electrical charge. Apart from hydrogen atoms all atomic nuclei contain neutrons as well as **protons**.

The neutron possess a mass (m_n) of 1.674 954 3 $\times$ 10^{-27} kg, which is about 0.1 per cent greater than the mass of the proton.

Nichrome source (i.r.). Basically a nickel and chromium alloy wire, this source can be unsupported or wound on a ceramic support. The wire itself radiates sufficient light of infrared frequency for it to have been widely used in infrared spectrophotometers. *See also* **Globar**; **Nernst glower**; **Opperman source**.

Nier–Johnson mass analyser (m.s.). Unlike the **Mattauch–Herzog mass analyser** this particular system uses electric and magnetic analysers both of 90° and deflecting the ions in the same direction[205] (see Figure 35, page 127). All ions are brought to a single focal point by varying either the accelerating voltage or the magnetic field. By using a magnetic sector of 30 cm radius and an electrostatic sector radius of 38 cm a resolution of 200 000 is possible.[206] *See* **Resolution: mass spectrometry**.

Nitrogen rule (m.s.). A substance containing an odd number of nitrogen atoms will have a molecular ion with an uneven mass number if the only other elements present are carbon, hydrogen, oxygen, sulphur or halogens. Similarly a molecule of even numbered molecular weight contains either no nitrogen or an even number of nitrogen atoms.

Non-bonding electrons. Electrons in the outer orbits of an atom in a molecule which are not involved in bond formation. Such electrons are easily removed by the electron bombardment in the mass spectrometer ion source and play a major role in leading to fragmentation and rearrangment of ions. Electrons in this classification include the electrons on oxygen atoms in carbonyl groups.

$$> C = \ddot{\underset{\cdot}{O}}$$

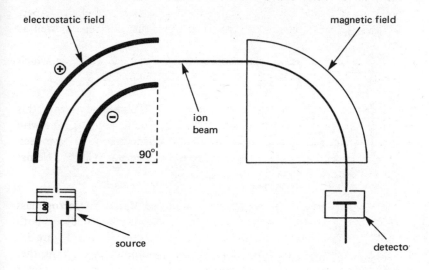

Figure 35. Nier–Johnson geometry for double focusing mass
spectrometer

Non-dispersive spectrometers (i.r.). Continuous monitoring of gases and
vapours is frequently carried out using spectrometers in which fixed
filters or prisms are employed to pass only a narrow radiation bandwidth. If the wavelength(s) passed correspond to a strong infrared
absorption band in a particular vapour the spectrometer can be used
as a sensitive selective detector for that substance. Such non-dispersive
spectrometers are of particular value in industrial environments for the
detection of toxic gases in an otherwise constant atmospheric background, one of the most widely used instruments being for the detection of carbon monoxide.

One special application of this type of spectrometer is in the detection of ethanol in the exhaled breath of drinking drivers.[207] The instrument used for this is shown diagrammatically in Figure 36. The filter is

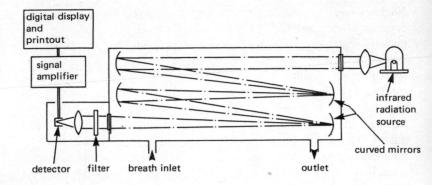

Figure 36. Infrared breath testing instrument employing multiple reflection

designed to detect the absorption at 3.48 μm as this is characteristic for ethanol and does not correspond to absorptions of most other substances found in breath.

Normal distribution. *See* **Gaussian curve (distribution).**

n to pi star and n to sigma star transitions. *See* **Bonding and antibonding orbitals** and **Bands (ultraviolet).**

Nuclear angular precession frequency (ω_N) (n.m.r.). The angular frequency for nuclear precession is related to the magnetogyric ratio and the applied field by the following equation

$$\omega_N = \frac{geB}{2m_p} = \gamma B$$

See also **Larmor frequency.**

Nuclear *g* factor (n.m.r.). *See* **Landé *g* factor**.

Nuclear magnetic resonance spectroscopy. A realm of spectroscopy concerned with the study of the interaction of energy with atomic nuclei. Most atomic nuclei possess a nuclear spin which can have a number of energy states depending upon the value of the nuclear **spin quantum number** for the nucleus. Nuclear magnetic resonance spectroscopy measures the energy necessary to bring about transitions between these energy levels by subjecting the nuclei to a powerful magnetic field and simultaneously irradiating it with a radio-frequency source.[208] In the case of protons suitable conditions for resonance occur with the use of a field of 1.4 tesla and a radio frequency of about 60 MHz (Figure 37, page 130). *See also* **Coupling constant** and **Spin–spin coupling**.

Nuclear magneton (μ_N) (n.m.r.). The unit for the measurement of nuclear magnetic moments; it has the value of $5.050\,824 \times 10^{-27}$ $J\,T^{-1}$.

$$\mu_N = \frac{m_e}{m_p}\,\mu_B = \frac{eh}{4\pi m_p}$$

Nuclear quadrupole moment (Q). Nuclei with spin quantum numbers greater than ½ have quadrupole moments due to the non-uniform distribution of the electric charge throughout the nucleus. Nuclear quadrupole resonance spectroscopy is used to study the transitions that can occur between the energy states that exist owing to this lack of symmetry.[209] Interaction with nuclear quadrupole moments leads to line broadening in n.m.r. spectra; this occurs particularly in the study of proton spectra when the protons are attached to nitrogen atoms ($I = 1$).

Nuclear quadrupole resonance. Owing to the existence of the **nuclear quadrupole moment** identical nuclei in different chemical compounds will exist at different energy levels. As a result they will give resonance peaks at different frequencies and this can serve as a guide to the nature

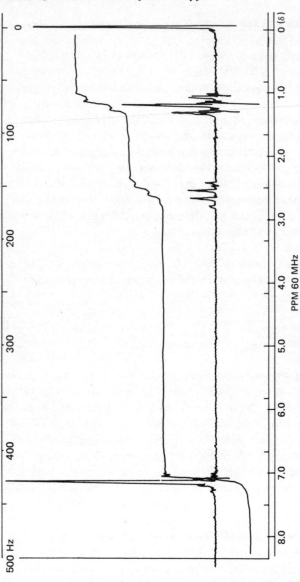

Figure 37. Proton magnetic resonance spectrum of ethyl benzene

of the bonding and the molecular structure. The effect is studied in nuclear quadrupole resonance spectroscopy,[210] the transitions occurring between 10^8 and 10^9 Hz.

Nuclear satellite signals (n.m.r.). Owing to the presence of 1% ^{13}C in naturally occurring carbon and its compounds all resonance spectra of protons attached to carbon atoms show small additional peaks due to **spin–spin coupling** between the hydrogen nuclei and the ^{13}C nuclei. These satellite signals are found on either side of the normal proton absorptions and the coupling constant J_{CH} is about 200 Hz. These satellite signals should not be confused with **spinning side bands**; their presence can be emphasized and useful information obtained by expanding the normal n.m.r. signals off-scale.

Nuclear screening constant (σ) (n.m.r.). The local magnetic field around a nucleus situated in the magnetic field of an n.m.r. spectrometer is usually less than the applied external field and is given by the equation

$$H_{local} = (1 - \sigma) H_0$$

where σ is the nuclear screening constant. This is the total of a number of diamagnetic and paramagnetic effects involving the circulation of electrons around the nucleus or nuclei studied and the production of small magnetic fields in opposition to the main field. *See* **Diamagnetism** and **Paramagnetism**.

Nuclear spin (n.m.r.). The nuclear spin of magnetic nuclei results in the **magnetic moment**; such nuclei possess a **spin quantum number** that determines the number of orientations that the spinning nucleus may occupy in the presence of an external magnetic field. Non-magnetic nuclei do not exhibit nuclear spin.

Nuclear spin quantum number (I) (n.m.r.). *See* **Spin quantum number**.

Nuclear spin temperature (T_s) (n.m.r.). An expression used to describe the number of nuclei in the different spin states. For protons

$$T_s = \frac{2\mu H}{k \ln(n_+/n_-)}$$

where n_+ and n_- are the number of nuclei in the upper and lower states respectively.

The spin temperature has a positive value when the population in the lower state exceeds that in the upper state.

Nujol (i.r.). A trade name for a high-boiling petroleum fraction used as a mulling medium in the study of solids by infrared spectroscopy. Being a mixture of hydrocarbons the only absorptions present are those due to C–C and C–H bonds occurring in four short regions: 3000–2700 cm^{-1} (3.3–3.6 μm), 1470–1410 cm^{-1} (6.8–7.1 μm), 1380–1350 cm^{-1} (7.3–7.4 μm) and 730–710 cm^{-1} (13.7–14.1 μm). *See also* **Mulls**.

O

Opperman source (i.r.). A ceramic (alkali based) tube with an internal noble metal heater. This type of source needs no preheating and emits powerfully at infrared frequencies immediately after being switched on. Commercially it is the cheapest and one of the most widely used radiation sources. *See also* **Globar**, **Nernst glower**; **Nichrome source**.

Optical comb. *See* **Attenuator**.

Optical density. *See* **Absorbance**.

Optical glass. Two main groups of optical glass are used; these are crown glass with a refractive index ≈ 1.52 and flint glass with a refractive index of about 1.65. The two types of glass are frequently used together to produce achromatic systems (*see* **Achromatic lens**).

Optical null principle (i.r.). The basis of this principle is the use of a servo operated **attenuator**. Typically, absorption of energy in the sample beam of a double beam system is matched, through the action of a servomotor, by the movement of an attenuator in the reference beam. The servo circuits are designed to drive the attenuator to the 'matched' position where there is 'null' difference between the energy transmitted by the two beams. The recorder pen follows the movement of the attenuator, thus providing the spectra.

Optical rotary dispersion (u.v.). The study of the variation of the angle of polarization for optically active materials with the change in wavelength. The effect is normally studied over the 200–700 nm spectral range using light monochromated from a xenon light source. It is used to differentiate between structures with different conformations and to establish the position of substituents in large molecules such as steroids.[211]

Ordinate. The vertical axis in a set of two-dimensional coordinates.

Usually this is referred to as the Y axis, so that the ordinate of any point in the plane of two axes forming the coordinates is the value at the place on the Y axis at which a horizontal line from the point meets that axis. *See also* **Abscissa**.

Overhauser effect (Nuclear electron double resonance) (e.s.r., n.m.r.). The study of nuclear resonance absorption signals while the sample is simultaneously irradiated at an electron resonance. The n.m.r. absorption signals for the spinning nuclei can be substantially enhanced by this irradiation if saturation of the associated spinning elections is obtained. This is because the equilibrium of the coupling between the spinning electrons and the spinning nuclei is altered and an increased population difference between the nuclear energy levels is produced.[212]

The reverse effect to the Overhauser effect[213] is **ENDOR**. Although the Overhauser effect is of particularly value for nuclear–electron spin interactions, it has also been used for the study of nuclear–nuclear spin interactions.

Overtones (i.r.). The term overtone is used in a general sense to apply to any multiple of a given fundamental frequency v (i.e. $2v$, $3v$, etc). In infrared spectroscopy most overtones are found in the near-infrared region below 2.5 μm (beyond 4000 cm^{-1}). Such absorptions are much weaker than are the main absorption bands, but may still be of spectroscopic interest. Aromatic compounds, for example, show overtone absorptions in the 2000–1667 cm^{-1} (5-6 μm) region which are characteristic for the type of aromatic substitution.[214]

P

Parallel vibration (i.r.). A vibration of a molecule leading to a change in dipole moment in a direction parallel to the symmetry axis. Such a vibration is infrared active. *See also* **Perpendicular vibration**.

Paramagnetic broadening (n.m.r.). One cause of the broadening of n.m.r. absorption peaks can be the presence of paramagnetic species in the sample under investigation. These have the effect of reducing the **longitudinal relaxation** time due to the much greater magnetic field associated with the paramagnetic substances. Dissolved oxygen is frequently a cause of paramagnetic broadening and free radicals can also have this effect.

Paramagnetism. A property of substances arising from the unbalanced spin of electrons around nuclei in molecules possessing unpaired electrons. Within an applied magnetic field the atoms tend to orientate themselves in the direction of the field and possess a positive magnetic susceptibility.[215] The degree of orientation parallel to the applied field increases with the strength of the field.

Parent ion (m.s.). *See* **Metastable ions**.

Parent peak (m.s.). The **molecular ion** formed by loss of a single electron from the molecule gives rise to the peak on the mass spectrum corresponding to the molecular weight. This peak is still sometimes referred to as being the 'parent peak' and has traditionally been indicated on the **bar graph** by the letter P. It is now more common for the molecular ion to be designated by the letter M, and the 'parent peak' should not be confused with the 'parent ion' used to refer to the precursor ion in a metstable transition (*see* **Metastable ions**).

Parts per million (p.p.m.) (n.m.r.). A dimensionless scale used on n.m.r. spectra in order that **chemical shift** positions relative to a standard can be expressed independently from the field strength of the

135

instrument employed.

For a particular absorption the value in p.p.m. is given by

$$\delta = \frac{(H_{ref} - H_{sample})\, 10^6}{H_{ref}}$$

In proton spectra δ values usually increase in a downfield direction away from the **TMS** reference signal; most absorption values lie between 0 and 10δ. Another system of numbering that refers specifically to TMS and protons is the τ scale, this numbers in the reverse direction so that $\tau = 10 - \delta$.

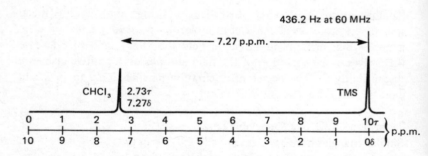

Figure 38. Parts per million. Proton n.m.r. spectrum of chloroform showing relationship between δ and τ parts per million and instrument frequency

PAS. *See* **Photoacoustic spectroscopy**.

Pascal (Pa). *See* **Pressure, units**.

Pascal constants (n.m.r.). In calculations of magnetic moments it is necessary to compensate for any contributing diamagnetic effects. The molar diamagnetic susceptibility (χ_m) is estimated from the sum of the atomic contributions, the values of which are known as Pascal constants.[216]

Pascal triangle (e.s.r., n.m.r.). In the **binomial series** $(x + y)^a$, successive coefficients can be obtained by forming a triangle of numbers such that a number in a line is obtained from the sum of the two numbers above it in the preceding line. In its simplest form the triangle serves to show the relative sizes of the peaks in multiplicities in first-order e.s.r. and n.m.r. spectra for nuclei in which $I = \frac{1}{2}$; as shown in table 6 below.

Table 6

Number of coupling nuclei	Ratio of multiplicity peak heights												
0							1						
1						1		1					
2					1		2		1				
3				1		3		3		1			
4			1		4		6		4		1		
5		1		5		10		10		5		1	
6	1		6		15		20		15		6		1

Similar triangles have been developed for $I = 1$ and $I = 3/2$.

Perfluorokerosene (m.s.). Because of the multiplicity of fragment ions obtainable from this fluorinated hydrocarbon it is used extensively as a standard for peak assignment in mass spectra. In some cases it is employed as an internal standard to produce an overlapping spectrum with the other material under examination. It is also used as a mulling agent, under the name **Fluorolube** for infrared spectra.

Permeability (μ). The ratio of the **magnetic flux density** passing through that body to the flux density of the magnetic field in the absence of the body.

$$\mu = \frac{B}{H}$$

It is measured in henry per meter, where $1 \text{ H m}^{-1} = 1 \text{ m kg s}^{-2} \text{ A}^{-2}$.

Perpendicular vibration (i.r.). A vibration of a molecule leading to a change in dipole moment in a direction perpendicular to the symmetry axis. Such a vibration will be infrared active. *See also* **Parallel vibration.**

Phosphorescence (pl.). Transitions from **triplet states** to the ground electronic state give rise to phosphorescent emissions. The lifetime of the triplet state is typically milliseconds but can be as long as ten seconds. Because of this long lifetime molecules in the triplet state may lose their energy by other deactivation processes. Phosphorescence is, therefore, best observed in solid solution of low temperature with the exclusion of oxygen.

Photoacoustic spectroscopy (PAS). Although PAS was first suggested more than 100 years ago[217] it is only recently that it has attracted any great attention. It is the study of the non-radiative de-excitation processes which take place after samples have absorbed radiation and the release of this absorbed energy by thermal transfer.[218] The theory of the subject[219] is based upon the concept that the modulated radiation absorbed by a sample causes that sample to be heated and cooled at the same frequency as that of the modulation. This regular temperature change is transmitted to any adjacent gas layer which expands and contracts with the variation producing an acoustic signal which can be detected and measured. The spectrum is obtained by measuring this signal as the radiation wavelength is varied.[220]

Photocell. Any cell made from photo-sensitive materials such that incident radiation causes the release of electrons that can be conducted away, amplified and recorded. Typical substances employed for this purpose are cadmium sulphide, lead sulphide and germanium. Lead sulphide has been of particular value in the development of the photoelectric infrared detectors and cadmium sulphide and selenide have been similarly used for visible light.

Photoelectric infrared detectors (lead sulphide conductive cells) (i.r.). Detectors suitable for a limited region of the near- and middle-infrared spectrum between 1 and 6 μm. They exhibit an increase in conductivity when infrared radiation is incident upon the surface, composed of lead sulphide, selenide or telluride.[221]

The detector consists of a 0.1 μm layer of the lead compound upon glass forming a semiconductor system which becomes a conducting system owing to the absorption of infrared radiation. The resistance of the detector decreases with the increase in intensity of the incident radiation.

Photoelectron spectroscopy (pe.). If substances are subjected to irradiation by short wavelength electromagnetic radiation (corresponding to high energy photons) they are found to emit electrons. This is called the photoelectron effect or photoionization and forms the basis of photoelectron spectroscopy.[222] This realm of spectroscopy is concerned with measurement of the kinetic energy distribution of the photoelectrons as these are a guide to the electronic structure of the material being irradiated. The photoionization may be brought about by wavelengths in the far-ultraviolet (vacuum u.v.) region and the use of **X-rays**. Standard gas discharge lamp u.v. source are normally sufficient to release the outer valence electrons, but X-ray energies are necessary to ionize the core electrons.[223]

Photoemissive tube (u.v.). A detector consisting of a photoemissive cathode, which emits electrons when radiation is incident upon its surface, and an anode collector enclosed in a glass jacket. Most commonly the cathode consists of caesium oxide, caesium, silver oxide and silver in a composite layer on a nickel base. Electrons released from the cathode due to incident radiation pass to the anode and the signal is amplified. Most photoemissive tubes give a good response over a range of about 300 nm.

Photoionization source (m.s.). Sample ionization for mass spectrometry can be achieved by irradiating the sample with ultraviolet light. The ultraviolet radiation source is separated from the ionization chamber by a lithium fluoride window. The method is limited to the formation of molecular ions and fragments formed by energies less than the 11.8 eV effective power from the light source used.

Photoionization spectroscopy (pe.). Although related to photoelectron spectroscopy this technique differs from it in the sense that the total photoelectron current from the irradiated material is measured without using a kinetic energy analyser. The photoelectron current is recorded as the frequency of the incident radiation is progressively increased from an initial level below that necessary for photoionization to occur.

Photoluminescence (pl.). The irregular re-emission of radiation from a molecule which has absorbed **electromagnetic radiation** from the infrared, visible or ultraviolet regions. It includes both **fluorescence** and **phosphorence**. Over a time period the amount of re-emitted radiation is equal in all directions.

Photometer. Traditionally, a word applied to any instrument used to measure the total amount of radiant energy absorbed or transmitted by a sample. A simple photometer does not, therefore, include a monochromator or a system for scanning a spectral range. *See also* **Spectrophotometer**.

Photomultiplier (a.s., u.v.). A sensitive, rapidly responding detector system in which photoelectrons produced by radiation falling on a cathode are amplified by a **cascade process**. Each electron from the gallium arsenide cathode produces two or more electrons by impact with the surface of a **dynode** plate. These are amplified in turn by impact on further dynode plates arranged in series (of which they may be more than ten). Because of the substantial amplification factor the photomultiplier is particularly useful in detecting signals of weak intensity to produce a measurable d.c. current.

Very sensitive photomultipliers were used in the ultraviolet spectrometers in the *Mariner 6* and *7* spacecrafts sent to Mars[224] in 1969. Caesium iodide was used as a photocathode for 110 to 210 nm and a bialkali metal for 190 to 430 nm.

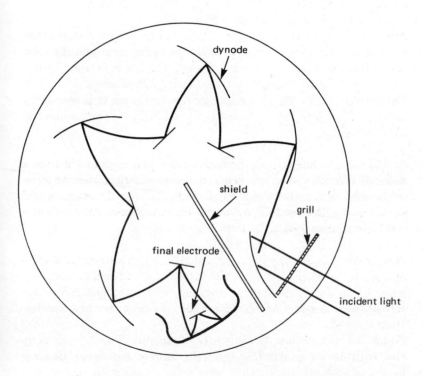

Figure 39. Cascade process in a photomultiplier

Photons. Electromagnetic energy is produced in discrete small quantities ('quanta') called photons. Photons produced at a particular frequency possess the same energy, but the amount of energy in a photon depends upon its frequency of radiation. The counting of photons is now a very important technique carried out using specially modified photomultiplier tubes which has improved methods for trace analysis in fluorimetry. *See* **Planck constant.**

Phototube. *See* **Photoemissive tube.**

Photovoltaic cell (u.v.). *See* **Barrier-layer cell.**

Pi to pi star transitions (u.v.). *See* **Bands (ultraviolet)** and **Bonding and anti-bonding orbitals.**

Planck constant (h). The proportionally constant in the equation

$$E = h\nu$$

derived for the statement of the Planck law that energy transfers occur with radiation composed of quanta of energy related to the frequency of the corresponding radiation

The value of h is $6.626\ 176 \times 10^{-34}$ J s.

It is sometimes used in the form h bar where

$$\hbar = \frac{h}{2\pi} = 1.054\ 5887 \times 10^{-34} \text{ J s}$$

It is also used as $h/2e = 2.067\ 8506 \times 10^{-10}$ J s C^{-1}.

Planck law of radiation. The intensity of radiation from a black body at a particular wavelength (the spectral radiant energy density) is given by

$$w_\lambda = \frac{8\pi hc\lambda^{-5}}{e^{he/kT\lambda} - 1}$$

All other equations concerning black-body radiation can be derived from the Planck equation. *See also* **Rayleigh and Jeans law; Stefan–Boltzmann law; Wien laws.**

Plane of symmetry (i.r.). A plane of symmetry is said to exist in a molecule if it is possible to draw a plane through the molecule in such a way that the half of the molecule on one side is a mirror image of the half on the other side. *See also* **Axis of symmetry** and **Centre of symmetry**.

Plasma jet (plasma torch) (a.s.). *See* **Inductive coupled plasma**.

Pneumatic cell (i.r.). Of the various detectors available for infrared radiation this is the most sensitive that has so far been developed.[225] The incident radiation causes the expansion of a small volume of gas contained in a closed cell. The expansion displaces a small mirror and reduces the amount of light reflected onto a photocell through a line grid.[226] The change in light intensity as a result of this displacement is proportional to the incident infrared radiation and is measured and recorded. The detector has been of special application in far-infrared spectrometers. *See also* **Bolometer** and **Thermocouple**.

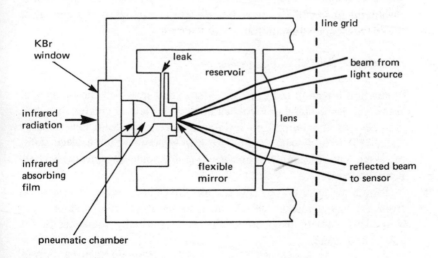

Figure 40. Pneumatic cell (Golay detector)

Polarizers. The examination of samples under polarized infrared radiation is of particular interest in the study of crystals and polymers. Several types of polarizers are made for this purpose. Stacks of cellulose lacquer films coated with selenium have been employed,[227] which will produce 99.8% polarization when inclined to the incident beam at the **Brewster angle**. Silver chloride plates 0.005 cm thick have also been employed in this way.[228]

Wire grid polarizers[229] are now the most commonly used. They consist of a network of fine parallel metal wires deposited on an infrared-transparent substrate. The metal is usually gold, while the substrate can be silver bromide (for 2.5 to 40 μm, with 2880 lines mm^{-1}), polyethylene (for > 15 μm, with 640 lines mm^{-1}),[230] or Irtran-2 and **Irtran-4**.[231]

Polychromatic radiation. Although literally meaning 'many coloured' the expression is used in connection with any electromagnetic radiation consisting of a mixture of wavelengths. Most sources in spectroscopy give polychromatic radiation which must be passed through a **monochromator** to obtain a selected wavelength or narrow band required for the spectroscopic examination.

Polystyrene. *See* **Calibration peaks**.

Population inversion (n.m.r., r.s.). A situation in which the number of active species occupying a particular energy state is greater than that predicted by the **Boltzmann ratio** and exceeds that of a lower energy state. This is of special consideration in **lasers** in which excited states of extended lifetimes are produced and population inversions deliberately created.

In n.m.r. spectroscopy population inversions occur if **relaxation** from the higher nuclear energy state to the lower energy state is too slow. This leads to band broadening and eventually total loss of the absorption signal.

Positron. *See* **Electron**.

Precession (n.m.r.). Nuclei possessing magnetic moments act as if they are small spinning magnets and will rotate when under the influence of an applied magnetic field. This motion is known as precession and takes place at a rate defined by the **Larmor frequency**. The precession axis of the spinning nucleus describes a circle at right angles to the direction of the applied field.

Pressed discs (i.r.). One method of obtaining infrared spectra from solids is by grinding and compressing a small quantity (*ca* 2 mg) of sample with dry potassium bromide or potassium chloride (*ca* 200 mg) under a pressure of 10 to 15 ton in^{-2} (154 to 231 MN m^{-2}). This method of sampling has the advantage that the solid supports employed do not absorb between 4000 and 700 cm^{-1}. The discs produced by this method measure about 1 cm in diameter and 1 to 2 mm in thickness and are normally reasonably transparent if the mixture has been adequately mixed and ground together.

Pressure, units (m.s.). The recommended **SI unit** for pressure is the pascal (Pa) measured in kg m^{-1} s^{-2} = N m^{-2} = J m^{-3}.

It has been suggested that although the bar is a multiple of the pascal its use should be progressively abandoned.

$$1 \text{ bar} = 10^5 \text{ Pa}$$

It has also been recommended that the use of the torr and millimetres of mercury should be discontinued.

$$1 \text{ torr} \approx 1 \text{ mm Hg} \approx 133.322 \text{ Pa}$$

Pressure broadening (a.s., em.). Broadening of absorption and emission lines in atomic spectra that occurs as a result of collisions between excited atoms and other atoms and molecules in the flame is called pressure broadening. This effect is partially overcome in atomic absorption spectroscopy by the use of low pressures in **electrodeless discharge tubes** and **hollow cathode lamps**. *See also* **Doppler broadening**.

Primary filter (pi.). The primary filter system is used to select the appropriate wavelength from the light source employed in the filter fluorimeter. The **filter** used may be of special glass or a chemical solution.[232] For the mercury arc source cobalt sulphate solution may be used to isolate the 253.7 nm line and a glass filter to obtain the 334.1 nm line.

Prisms, infrared. The prisms used for dispersion in infrared spectrometers are of many types as no prism is completely suited for the entire infrared region. Most general purpose prism instruments employ a sodium chloride prism, but operation in the far infrared necessitates the use of other materials. The following table gives the operating ranges for the most common prism materials. *See also* **Gratings**.

Table 7

Material	Frequency range/cm^{-1}	Wavelength range/μm
caesium iodide	5000–200	2.0–50.0
sodium chloride	5000–650	2.0–15.4
KRS-5	5000–1300	2.0– 7.7
calcium fluoride	4500–1300	2.2– 7.7
lithium fluoride	4500–1700	2.5– 5.9
potassium bromide	1100–285	9.1–35.0

Prisms, ultraviolet. Although glass is an admirable material for prisms for studying visible spectra it is unsuitable for the ultraviolet region. For this reason most modern general-purpose prism ultraviolet spectrometers employ a quartz or silica prism capable of operating over a wider spectral range. The normal operating limits are shown below

Table 8

Material	Use range/nm
glass	350–1000
quartz	200–1000

Probes (i) (m.s.). *See* **Direct insertion probes.**

Probes (ii) (n.m.r.). The part of the n.m.r. spectroscopy which is situated between the pole faces of the magnet. It includes the sample chamber which enables the sample tube to be positioned precisely with respect to the irradiation and detection coils constained in the probe. The probe itself has to be symmetrically positioned in the homogeneous portion of the magnetic field of the instrument. Because it is virtually a self-contained section a single n.m.r. instrument may be supplied with several interchangeable probes each suitable for the study of a different element.

Proton. Of the three fundamental particles making up atoms, the proton is the only one possessing a positive charge. The proton, which can be considered as the nucleus of a hydrogen atom possesses a mass (m_p) slightly less than that of the **neutron** of $1.672\,648\,5 \times 10^{-27}$ kg and is 1836 times greater than that for the electron.

Pulsed laser (r.s.). *See* **Laser.**

Pulsed mode (m.s.). Operation of a pulsed-mode time-of-flight mass spectrometer involves ionization of the sample by short pulses of the electron beam (250 ns). The positive ions formed are then released into the accelerating region by a negative pulse (300 V) lasting 2.5 μs on the ion focus grid. *See also* **Continuous mode.**

Pulse method (n.m.r.). Longitudinal and transverse **relaxation times** can be determined by pulse experiments in which the n.m.r. spectrum is studied by means of short bursts (short compared with T_1 and T_2) of radio-frequency energy at the resonance energy. The application of correctly timed amounts of energy leads to a nuclear signal (spin echo) whilst no radio-frequency energy is applied. The method has also been used to measure **coupling constants.**

Q

Quadrupole broadening (n.m.r.). The width of an n.m.r. absorption peak is broadened if there is a decrease in the **relaxation time** as can be caused by the existence of a non-spherical charge distribution on other or the same nuclei (*see* **Quadrupole moment**). This is known as quadrupole broadening and occurs in the presence of nuclei in which $I > \frac{1}{2}$ which possess quadrupole moments. This is the reason for the broad absorption signals obtained for protons attached to ^{14}N.

Quadrupole coupling (n.m.r.). The interaction of two nuclei, one of which possess a non-spherical charge distribution (*see* **Quadrupole moment**). It leads to a broadening of the n.m.r. absorption signal of the other magnetic nucleus. *See also* **Quadrupole broadening**.

Quadrupole mass spectrometer (m.s.). Mass spectra — up to a mass/charge ratio greater than 1000 — can be obtained by passing the ion beam along the centre axis of four parallel circular or hyperbolic quartz rods coated with platinum which are connected in pairs to radio-frequency and d.c. supplies. These voltages create a hyperbolic field between the rods and only one particular mass value can traverse the path between the rods for any specified combination of r.f. and d.c. voltages (see Figure 41, page 149). This system has the advantage that mass peaks are separated by a constant distance and do not crowd together at high mass values.[233,234]

Typical dimensions of the quartz rods in the mass quadrupole are about 12 mm diameter and 25 cm long, operating with 3 MHz r.f. and 10 V d.c. Smaller quadrupole mass spectrometers are used as residual gas analysers and simpler monopole mass spectrometers employ only a singe rod.

Quadrupole moment (Q) (n.m.r.). Nuclei which have a **spin quantum number** $I > \frac{1}{2}$ possess electric quadrupole moments as a result of a non-spherical distribution of the charge in the nucleus. Nuclei in which $I = \frac{1}{2}$ cannot possess electric quadrupole moments. Those nuclei

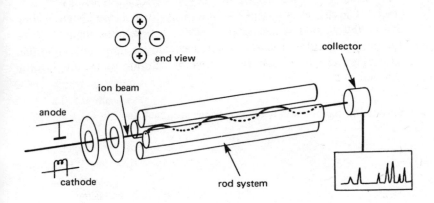

Figure 41. Quadrupole mass spectrometer

possessing quadrupole moments, such as ^{2}H, ^{11}B and ^{14}N, cause **quadrupole broadening** of n.m.r. absorption signals from other nuclei.

Quanta. *See* **Photons.**

Quantum numbers. Numerical values used to specify the size and shape of orbitals and/or to govern the energy levels of electron and nuclear spin systems. The four atomic quantum numbers for electrons are

Table 9

Quantum number	symbol	defining	permitted value
principal	n	orbital radius	integral from 1
secondary	L	angle	integral from 0 to n − 1
magnetic	M	angle	integral from 0 to ± L
spin	S	spin	± ½

Quarter-wave plate. *See* **Retardation plate.**

Quenching (pl.). The process whereby fluorescence is decreased as a result of a change in pH of solution or by some added chemical that assists the excited molecules to transfer their excess energy before fluorescent emission can take place. Quenching as a result of impurities and added chemicals can be reduced by dilution of the fluorescent solution.

R

Radiant energy. Energy which is transmitted by means of electromagnetic waves without involving the transfer of matter.

Radical ions (m.s.). Removal of an electron from a molecule due to bombardment with electrons from the ion source leads to the formation of a positively charged radical ion. This radical ion is detected on the mass spectrometer and recorded as the molecular ion

$$R : R' \xrightarrow{\ -e\ } R \overset{+}{\underset{\cdot}{}} R'$$
$$\text{radical ion}$$

Radio-frequency spark discharge source (m.s.). *See* **Spark ionization source**.

Raies Ultimes (em.). *See* **R.U.**

Raman effect (r.s.). Although originally predicted by Smekal[235] this effect was first examined by Raman[236] in 1928. He observed that a small proportion of radiation passing through a substance emerged with either an increase or a decrease in the frequency. This occurred because after being raised to a quasi-excited (or virtual) state, following inelastic collisions with photons, some of the excited molecules returned to a higher or lower vibrational level in the ground electronic state. Raman transitions may also be vibrational–rotational or simply rotational[237] in nature. The lifetime of the Raman excited state is about 10^{-12} s and only about 1 in 10^6 molecules exhibit the effect at any particular moment. *See also* **Anti-Stokes lines** and **Stokes lines**.

Raman spectroscopy. Modern Raman spectroscopy is carried out with lasers as the source of radiation.[238] The spectra obtained are complementary to normal infrared spectra and assist in the determination of molecule dimensions. The Raman lines are observed on either side of the monochromatic laser frequency and are referred to as **Stokes lines**

and **Anti-Stokes lines**. Visible light is normally employed for the purposes of excitation as the intensity of the resulting Raman lines is proportional to the fourth power of the frequency of the irradiation.[239] *See also* **Laser Raman spectroscopy**.

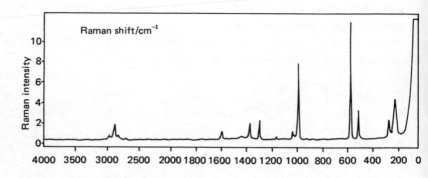

Figure 42. Raman spectrum of mesitylene

Rapid scanning spectrometer. A number of specially designed spectrometers capable of measuring spectra in fractions of a second. These are particularly suitable for studying combustions and rapid reactions in solution. The instruments employ a **pneumatic cell** detector with a cathode-ray oscilloscopy and a rapidly moving **Littrow mounting**.

Rayleigh and Jeans law. An attempt to develop an equation for blackbody radiation using statistical mechanics was used by Rayleigh and Jeans. The equation they obtained agrees with experimental values at long wavelengths and at high temperatures and is given by

$$M_\lambda = 2\pi hc^2 \lambda^{-5}(\lambda T)$$

See also **Planck law; Stefan–Boltzmann law; Wien laws**.

Rayleigh scattering (r.s.). Some light passing through a transparent medium is scattered instead of following the main light path. In the case of elastic (Rayleigh) scattering there is no change in the wavelength of the scattered light. Most of the scattered radiation is in this category, but a small amount may undergo inelastic scattering giving rise to the **Raman effect**. About 1 part in 10^3 of the incident light undergoes Rayleigh scattering and the amount of light scattered (I_θ) in a direction θ to the incident beam is given by

$$\frac{I_\theta r_s^2}{I_0} = \frac{8\pi^2 \alpha^2}{\lambda^4} \ (1 + \cos^2\theta)$$

R bands (u.v.). Absorption bands arising from n → π* transitions (from the German *radikalartig*) according to the system of nomenclature used by Burawoy[240] and Braude.[241] Such transitions occur in **auxochromes** possessing non-bonding electrons. *See* **Bands (ultraviolet)**.

Reaction field theory (n.m.r.). An attempt to explain the effect of solvents on **chemical shifts** in n.m.r. spectroscopy.[242] It concerns the effect of solvent dielectric constants and polar groups. The polar groups are said to set up 'reaction fields' in the solvent causing the electrons in the bonds to be redistributed, thus reinforcing the **shielding** or **deshielding** within the molecules.

Red shift (u.v.). *See* **Bathochromic shift**.

Reflection. A beam of radiant energy is said to be reflected when its line of direction is changed along another single direction as a result of meeting a smooth polished surface. When reflection occurs the angle between the incident beam and the surface is identical with that between the reflected beam and the surface. Also the two beams and the normal at the point of incidence are in the same plane.

Reflectometer (i.r., u.v.). Reflection spectra are obtained from the surfaces of samples fitted into reflectometer attachments for conventional spectrometers. This may be in the form of a diffuse reflectance unit such as that used in ultraviolet spectrometers for the study of papers, powders and tiles by comparison with a magnesium oxide standard. These give front-surface reflection spectra which are influenced by particle size. Reflectance spectra are also obtained by **attenuated total reflectance**.

Refraction. The change in direction which occurs when a beam of radiant energy passes from one medium into another of lesser or greater optical density. The angle of incidence of the beam and the angle of refraction, relative to a perpendicular to the two surfaces at the point of incidence, are related by the equation

$$\frac{\text{sine } i}{\text{sine } r} = \frac{n'}{n}$$

where n and n' are the refractive indices of the two media. In those cases in which the incident ray passes through air n is taken as unity and the equation becomes

$$\frac{\text{sine } i}{\text{sine } r} = n'$$

As the value of n' varies with both temperature and wavelength it is usual to indicate both of these; n_D^{20} means that the refractive index was determined at $20\,^{\circ}\text{C}$ using the yellow sodium D lines at 589.0 and 589.6 nm.

Relaxation (n.m.r.). The process of change from a high nuclear energy level to a lower energy level. Relaxation may be of two types: **longitudinal relaxation** or **transverse relaxation**. In solids or viscous liquids the relaxation time may be hours, but in non-vicous liquids relaxation times vary from 10^{-9} to 10^9 s. Where relaxation is slow compared with the excitation process **saturation** takes place.

Resolution: mass spectrometry. The ability of the **mass analyser** to separate two ions of mass M and $M + \Delta M$ from each other, the resolution being given by

$$R = \frac{M}{\Delta M}$$

Considerable conflict exists over the manner in which data on resolution is presented as in some cases this refers to peaks which are only just distinguishable from each other, as when there is 90% overlap (valley) which is poor resolution; whilst in other cases it may refer to 50% valley or 10% valley. These are illustrated in the figure below. Most responsible manufacturers now specify the resolution as the mass value at which the 10% valley is obtained. Using this definition most general-purpose mass spectrometers of the single focusing type (*see* **Single-focusing mass spectrometers**) have resolution values below 2000, whilst double-focusing instruments have values greater than 10 000.

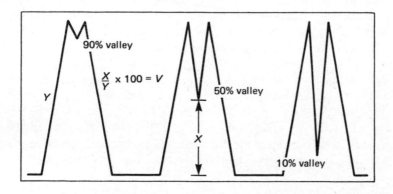

Figure 43. Resolution in mass spectrometers

Resolution: spectrophotometers. The ability of a monochromator to separate radiation of similar wavelengths is known as resolution and is given by the formula

$$R = \frac{\lambda}{\Delta\lambda}$$

when the two wavelengths λ and $\lambda + \Delta\lambda$ can just be differentiated from each other. Resolution can be improved by a narrow slit width and good dispersion.

The resolving power for a prism is

$$R = t \frac{\partial n}{\partial \lambda}$$

where t is the base width and $\partial n/\partial \lambda$ is known as the dispersive power of the prism.

Resolution in modern ultraviolet spectrophotometers is better than 0.2 nm at 250 nm, whilst in simple filter colorimeters it may be as poor as 10 nm at 500 nm.

Response time (i.r.). The time required by the detector in a spectro-photometers to reach equilibrium when there is a change in the amount of radiant energy falling on its sensitive surface. For modern detectors this is less than 0.03 s. *See also* **Time constant**.

Retardation plate. A phase difference is created between the ordinary and extraordinary rays of polarized light by passage of an incident ray through a retardation plate. The plates are formed from double-refract-ing materials cut parallel to the optic axis. The effect of a quarter-wave plate is to introduce a phase difference of 90° between the two emer-gent rays, and a half-wave plate introduces a phase difference of 180°.

Retarding field analyser (pe.). *See* **Electron energy analyser**.

Ringbom plots (u.v.). If the transmittance values for a series of colori-metric standards are plotted against the logarithm of the concentra-tion of the substance being measured an S shaped curve is obtained. This is a Ringbom plot[243] and the shape of the S is a guide to the suitability of the chromogenic reagent in terms of compliance with the

Beer–Lambert law and precision. Optimum concentrations for quantitative determinations are those falling on the central straight portion of the S line.

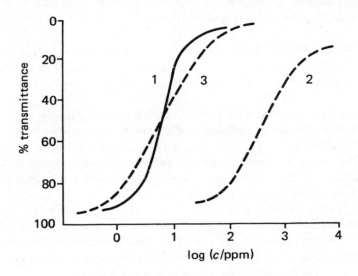

1. Sensitive, precise, narrow optimum range.
2. Suitable for higher concentrations only.
3. Less sensitive, wider optimum range.

Figure 44. Ringbom calibration curves

Ring current effect (n.m.r.). Aromatic protons are found to absorb at low field in n.m.r. spectra because of deshielding of the ring protons. The deshielding is attributed to the toroidal shape of the induced field due to the electron distribution above and below the plane of the benzene ring. This is called the ring current effect and occurs with other unsaturated systems.

Ringing (n.m.r.). The exponentially decaying oscillations which are frequently observed with sharp n.m.r. absorption signals. These are particularly associated with the **TMS** reference peak. It occurs at the high-field end of any sharp peak if a rapid sweep of the spectrum is carried out from the low- to the high-field direction. A good symmetrical ringing oscillation on the TMS absorption is usually taken as an indication of an homogeneous magnetic field.

Rocking vibration (i.r.). One of the forms of **bending vibration** in which the vibrational motion of the group is in the form of a rocking movement within the plane of the group.

Rotating sector mirror (i.r.). A circular mirror with alternate segments cut away so that on rotation it will alternately reflect and pass any light directed to it. The system is frequently used in spectrophotometers when it is required to pass two beams along the same path in alternate pulses. In this application one of the beams comes from behind the rotating mirror and passes through when in line with the cut-away portions and follows the same path as the beam of incident light on the reflecting segments of the mirror.

Rotational spectra (i.r.). Pure rotational spectra are found in the far-infrared region (*see* **Infrared spectra**) between 50 μm and 200 μm and in the microwave region (*see* **Microwaves**). They arise from the absorption of electromagnetic radiation and its total conversion to produce or change the molecular rotation. In the middle-infrared region rotational changes occur simultaneously with the vibrational changes (*see* **Vibrational spectra**).

Rowland mounting (em.). A slit, a grating and a photographic plate are arranged at the three points of a right-angled triangle in such a way that the hypotenuse of the triangle is the diameter of a circle passing through the three points (see Figure 45, page 159). This arrangement is used in some emission spectrographs in which the spectrum is scanned by moving both the grating and the photographic plates. The Rowland mounting is also of considerable importance in X-ray photoelectron spectroscopy where accurate focusing of monochromated X-rays is required.

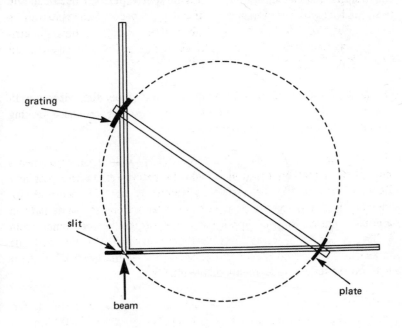

Figure 45. Rowland mounting

R.U. (Raies Ultimes) (em.). The most persistent lines in the emission spectrum of any element. They are the lines which are the last to disappear as the concentration of that element is progressively decreased. Because of this they are of great importance for detecting trace amounts of the particular element.

Specially manufactured R.U. powders are mixtures of substances employing concentrations that will only give the R.U. lines of the elements incorporated and are used as reference standards and for calibration purposes.

Rydberg constant (R_∞). The Rydberg constant (now expressed in terms of waves per metre) was introduced as the constant required for the equation for the calculation of the **atomic spectra** series. It is calculated from the following relationship:

$$R_\infty = \frac{\mu_0{}^2 m_e e^4 c^3}{8h^3} = 1.097\ 373\ 177 \times 10^7\ \text{m}^{-1}$$

in the formula for spectra of hydrogen and other light atoms and ions the value of R_∞ is modified by allowing for the mass of the nucleus (M) according to the equation

$$R_H = R_\infty \left(\frac{M}{M + m_e}\right)$$

S

Saturation (n.m.r.). When the number of nuclei occupying a high energy state and a low energy state are equal, saturation is said to have taken place. This leads to a loss of the n.m.r. absorption peak, which progressively broadens and decreases in intensity as saturation is approached. This situation can be treated by the application of a powerful radio-frequency signal. The attainment of saturation is greatly dependent upon the speed with which nuclei in the higher energy state can revert to the lower energy state. In liquids this rate of **relaxation** is rapid, but in solids the excited state might persist for several minutes or even hours. *See also* **Longitudinal relaxation** and **Transverse relaxation**.

Saturation factor (Z_0) (n.m.r.). A mathematical factor defining the degree by which absorption signals are diminished under conditions which tend to cause equalization of populations of nuclei in the different **spin states**. For nuclei possessing a spin quantum number $I = \frac{1}{2}$ and in a radiofrequency field H_1

$$Z_0 = \frac{1}{1 + \gamma^2 H_1^2 T_1 T_2}$$

Scattering. When radiation is incident on a surface or passes through a transparent material a proportion of the radiation is scattered in all directions. The nature of the scattering is dependent upon the wavelength of the radiation, the nature of the substance causing the scattering and upon inhomogeneities in the material. *See also* **Rayleigh scattering**.

Schrödinger wave equation. This equation is the basis of wave mechanics and is obeyed by any moving particle.

$$\frac{\partial^2 \Psi}{\partial x^2} + \frac{\partial^2 \Psi}{\partial y^2} + \frac{\partial^2 \Psi}{\partial z^2} + \frac{8\pi^2 m}{h^2}(E - V)\Psi = 0$$

161

Solutions of the equation are only possible for finite, non-zero, unique values of E; such solutions are termed **eigenvalues** and may be compared to the energy states developed from the Bohr theory of the atom. Ψ is termed the **eigenfunction** or wavefunction of the electron, and the relative probability of finding the electron at a particular point defined by x, y and z is given by Ψ^2.

Scissor vibration (i.r.). A form of **bending vibration** in which the motion of the atoms leads to a change in the angle between the atoms forming the group.

Screening constant (n.m.r.). *See* **Nuclear screening constant** and **Shielding**.

Secondary filter (pl.). A filter used in filter fluorimeters to absorb any reflected or scattered exciting radiation and to permit passage of only the fluorescent light. The **filters** are made from special glasses chosen to cut off radiation below a defined wavelength. A series of filters covering cut-off limits over the range 200–600 nm is available for use in fluorimeters.

Second-order (and higher-order) spin patterns (n.m.r.). When first-order spin conditions no longer apply (approximately when $\Delta \nu < 6 J$; *see* **First-order spin pattern**) the nature of multiplicities from **spin-spin coupling** becomes increasingly more complex. Thus the spacing and ratios of peak heights become distorted and frequently extra peaks occur. Such systems are termed second order (or occasionally a higher order than this).

Simplification of spectra and the determination of the coupling nuclei can be achieved by **multiple resonance** techniques and by the use of **chemical shift reagents**.

Sector disc (em.). *See* **Log sector** and **Step sector**.

Selection rules for atomic spectra. Certain transitions between electronic energy levels have been found to be almost totally forbidden in the formation of **atomic spectra**. The selection rules enable the permitted transitions to be predicted on the basis of the differences between the quantum number for any two energy levels involved.

For permitted transitions

ΔL (change in orbital angular momentum) = ± 1

ΔJ (change in total angular momentum) = 0 or ± 1

ΔM (change in magnetic orientation) = 0 or ± 1

ΔS (change in spin orientation) = 0

Shielding (n.m.r.). Nuclear magnetic resonance absorptions occurring at high field strength do so because of the high electron density around the nuclei. They are said to be shielded and this is assisted by adjacent electropositive groups and atoms. The circulating electrons around the nuclei incude a magnetic field that is opposite to the direction of the applied magnetic field. The effective magnetic field experienced by a nucleus as a result of shielding is $H_0 - \sigma H_0$, where σH_0 is the induced field due to the electron circulation. The larger the value of σH_0, the greater is the degree of shielding and the higher the applied field necessary to bring about resonance of the nucleus (i.e. in the case of protons the nearer this will be to the TMS signal). *See also* **Deshielding** and **Ring current effect**.

Shift reagents (n.m.r.). *See* **Chemical shift reagents**.

Shim coils (Golay coils) (n.m.r.). In the n.m.r. spectrometer it is necessary to be able to compensate for any small variation in the main magnetic field.[244] This is achieved by placing conducing coils, the shim coils, at the pole faces capable of carrying an electric current that will lead to the production of additional small fields to help maintain the uniformity of the field around the sample.[245] These coils measure about 2.5 cm in diameter.

They are made by designing conducting coils as patterns formed by etching thin copper sheet and placing them in sets of matching pairs, one from each set on each pole face, so that current carried by them will produce field effects around the sample. In practice a set of these shim coils may consist of up to nine different patterned pairs cemented together in two very thin sandwiches of copper sheets with insulation between.

Shpol'skii effect (pl.). Low-temperature fluorescent spectra (< 100 K) from materials in solution frequently show improved resolution of the vibrational bands compared to spectra obtained at room temperatures; this is known as the Shpol'skii effect.[246] The quality of the low-temperature spectra is solvent dependent, the best results being obtained when the dimensions of solvent and solute molecules are approximately equal. It has found particular application in the detection and determination of polynuclear hydrocarbons.[247] *See also* **EPA mixed solvent**.

Side bands (n.m.r.). *See* **Nuclear satellite signals** and **Spinning side bands**.

Sigma to sigma star transitions (u.v.). *See* **Bonds (ultraviolet)** and **Bonding and antibonding orbitals**.

Signal to noise ratio (S/N ratio). A measure of the signal in a peak measured relative to the noise (the background signal) which can be applied in some ways to all spectroscopic and recording systems. As noise is a random effect, if it is averaged over a period of time it should produce an almost straight line, while any proper electronic signal averaged in the same way should stand more clearly resolved from the noise. This is the basis of the **computer of average transients**. The detection limit for an instrument is that point at which the signal for a determination cannot be readily discriminated from the noise level, this is frequently considered to be at a signal to noise ratio of 2.

Simple cleavage (m.s.). The fragmentation of a single covalent bond to produce an ion and a neutral particle. Because bonds vary in strength, fragmentation of a molecule occurs at certain preferential sites and leads to relatively simple mass spectra.

Single-coil method (n.m.r.). The bridge, or single-coil method of measuring n.m.r. absorptions uses a radio-frequency bridge system similar to the Wheastone bridge arrangement (*see* **Wheatstone bridge circuit**).[248,249] So that the bridge is balanced when no energy is absorbed by the sample. When absorption occurs the bridge is thrown out of balance and a signal recorded. This is achieved with just a single cell arranged around the sample tube. *See also* **Induction method**.

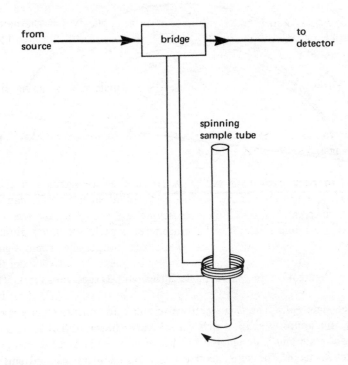

Figure 46. Single-coil probe

Single-focusing mass spectrometer (m.s.). A mass spectrometer possessing only one magnetic or electric field for resolving the positive ions. By far the most common of these are based upon magnetic analysers as originally developed by Dempster.[250] The **resolution** of such instruments is usually below 2000. *See also* **Double-focusing mass spectrometer.**

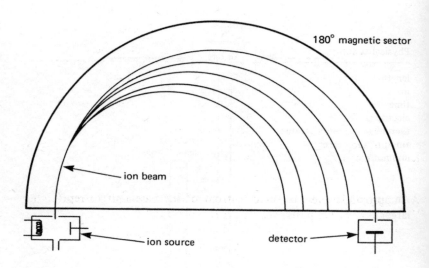

Figure 47. Principle of single focusing mass spectrometer

Singlet state (pl.). The two electrons in any filled energy level possess equal and opposite spins which cancel each other out. This is termed a singlet state and is designated by the letter S. When radiant energy is absorbed one of the two electrons is elevated to an excited singlet state S_1 which persists for a short period of time between 10^{-4} and 10^{-8} s. From the excited state three main processes are possible; the electron may revert to the ground state by loss of energy without re-emission to the surroundings, or it may pass to a **triplet state** eventually.

giving rise to phosphoresence, or the third process is loss of energy and reversion to the ground state via **fluorescence**.

SI units. A system of units based upon the metre, kilogram, second, ampere, kelvin, candela and mole, adopted by the Conference Générale des Poids et Mesures in 1960. This International System of Units (Système International d'Unités) employs only one basic unit for each physical quantity.[251,252,253]

Table 10

Physical quantity	SI unit	Symbol
length	metre	m
mass	kilogram	kg
time	second	s
electric current	ampere	A
thermodynamic temperature	kelvin	K
amount of substance	mole	mol
luminous intensity	candela	cd

An appropriate prefix is used in front of the base unit to represent larger and smaller units; these are listed in Table 11.

Table 11

fraction	prefix	symbol	multiple	prefix	symbol
10^{-1}	deci	d	10	deca	da
10^{-2}	centi	c	10^2	hecto	h
10^{-3}	milli	m	10^3	kilo	k
10^{-6}	micro	μ	10^6	mega	M
10^{-9}	nano	n	10^9	giga	G
10^{-12}	pico	p	10^{12}	tera	T
10^{-15}	femto	f	10^{15}	peta	P
10^{-18}	atto	a	10^{18}	exa	E

Slit. In the monochromation of light in spectrometers the size of the slit and its position is of considerable importance. The amount of radiant energy entering or emerging from the monochromator is determined by the slit width which needs to be varied with the wavelength to ensure roughly constant energy from the source over the whole spectral range. This adjustment is automatically carried out in most modern spectrometers in which the slit is coupled to the rotation of the **prism** or **grating**. Bilaterally symmetrical slits are now commonly used so that both sides move equally from the fixed central path of the radiation. Slit widths for infrared studies are between 10 and 1000 μm, whilst for ultraviolet work they are about 10 to 20 μm.

Smekal–Raman effect. *See* **Raman effect**.

Sodium-2,2-dimethyl-2-silapentane-5-sulphonate (n.m.r.). *See* **DSS**.

Soft-X-rays. *See* **X-rays**.

Solar blind detectors (u.v.). Special **photomultipliers** sensitive to middle- and vacuum-u.v. regions but 'blind' to solar radiation.[254] They incorporate a photo-cathode of either rubidium telluride or caesium iodide and **dynodes** of silver/magnesium.[255] The operating range is between 160 and 320 nm with a peak sensitivity at about 220 nm.

Solid deposits (i.r.). Infrared spectra of solids may be obtained from a thin film of the solid deposited on a sodium chloride or potassium bromide plate by evaporation of a small volume of a concentrated solution of the solid. This method produces an amorphous film sufficiently thin and transparent for the infrared radiation to pass through. The existence of large crystals leads to the production of poor spectra with broad absorption peaks. *See also* **Mulls** and **Pressed discs**.

Spark ionization source (m.s.). A radio-frequency spark source employing a voltage between 6 and 60 V is one method of obtaining suitable ionization to produce mass spectra of inorganic materials.[256] The frequency of the spark is 100 Hz with a pulse duration of 100 μs. The electrodes may be prepared from a mixture of graphite and the material to be examined.[257] Quantitative analysis of impurities as low as 10 p.p.b. can be carried out by this mass spectrometric procedure.

Spectral bandwidth. *See* **Effective bandwidth.**

Spectrofluorimeter (pl.). An automatic scanning instrument capable of studying fluorescence over a wide range of wavelengths. It consists of a radiation source (such as the xenon lamp), a monochromator for selection of the exciting wavelength, a sample cuvette and an emission monochromator with a photomultiplier detector. The analyser system is commonly at 90° to the source and excitation monochromator to minimize the effects of reflect and scattered radiation.

Spectrograph. An instrument which produces a spectrum and records it on a photographic plate. Early mass analysers were spectrographs, but the production of photographic images of spectra is now almost completely restricted to emission spectrography. Although it is the visible region of the spectrum that is normally examined by this means the use of special photographic plates has enabled it to be extended to the ultraviolet and near-infrared regions. Most spectrographs are prism instruments and employ spark sources, they are capable of detecting less than 1 part in 10^5 of most metals.

Spectrometer (optical). Any instrument which produces a spectrum by **dispersion** of radiant energy and in which measurement of the transmitted energy may be made by scanning the spectrum range. *See also* **Mass spectrometer** and **Spectrophotometer.**

Spectrophotometer. A highly sophisticated spectrometer for studying emitted radiation in the infrared, visible or ultraviolet regions, incorporating devices for the automatic measurement and recording of spectra. The essential parts to such an instrument are the source, the **monochromator**, the sample section, the detectors and the recorder. Any change of prisms or gratings during the course of scanning the spectrum is carried out automatically and the recorder provides a plot of the intensity transmitted or absorbed against the wavelength.

Spectroscopic splitting factor (e.s.r.). *See* **Landé *g* factor**.

Speed of light (*e*). Measured in a vacuum this has the value of 2.997 458 $\times 10^8$ m s^{-1}.

Spin echo (n.m.r.). *See* **Pulse method**.

Spin–lattice relaxation (T_1) (n.m.r.). *See* **Longitudinal relaxation**.

Spinning side bands (n.m.r.). Very small peaks in the background of n.m.r. spectra which are frequently observed on either side of a main absorption peak or **multiplet**. They arise from inhomogeneities in the permanent magnetic field and from lack of uniformity in the spinning sample tube. The measurement in hertz between the main n.m.r. absorption and the attendant spinning side bands is the rate of spin of the sample tube. Further spinning side bands can sometimes be observed at other multiples of the spinning rate.

Spinning side bands are frequently confused with other small peaks arising in the spectrum, such as those due to **nuclear satellite signals**. Their true nature can be ascertained by spinning the sample tube at different rates. The only bands that will move their positions relative to the stronger absorptions are the spinning side bands, while any true n.m.r. absorption signal will remain stationary.

Spin quantum number (*I*) (n.m.r.). Most nuclei possess a nuclear spin which may occupy several orientations in the presence of a magnetic field. The spin quantum number (*I*) defines the number of orientations ($2I + 1$) which the nuclear magnet may occupy in the applied magnetic field. *I* may be of integral or half integral value (e.g. ^{1}H, $I = \frac{1}{2}$; ^{11}B, $I = 3/2$; ^{14}N, $I = 1$). A few elements (those in which both the atomic number and the mass number are even) have $I = 0$ (e.g. ^{12}C and ^{16}O).

For nuclei in which $I = \frac{1}{2}$ there are two possible orientations in the applied field: either aligned with the field; or aligned against the field. The first of these three is the lower energy level.

Spin–spin coupling (n.m.r.). The influence that closely situated spinning nuclei have upon each other as a result of spin effects transferred through the bonding electrons. Spin–spin coupling occurs becasuse the bonding electron is influenced by the field produced owing to the spin of the nearest nucleus. The field of this electron will in turn affect the field of the second bonding electron which also continues the effect to the next nucleus. The overall result is that one nucleus has an indirect influence on the spin characteristics of another.

The effect diminishes with distance and rarely extends beyond four bonds between protons. Phosphorus nuclei, however, exibit long-range spin–spin coupling that frequently extends beyond four bonds.

The result of spin–spin coupling is that any particular nucleus is influenced by the different energy states (two in the case of protons) of any nearby nucleus. As a consequence its absorption peak is split into a **multiplet**. When coupling occurs with more than one proton the multiplicity of the resulting absorption can be determined from the **Pascal triangle**. *See also* **Coupling constant** and **Spin–spin decoupling**.

Spin–spin decoupling (n.m.r.). An instrumental procedure whereby the normal **spin–spin coupling** between nuclei is prevented by irradiating the two coupled nuclei each with a radio frequency related to the difference in the chemical shifts of the two nuclei. The technique used for this is known as **multiple resonance**, in its simplest form employing two radio frequencies it is double resonance. *See also* **INDOR** and **Tickling**.

Spin states (n.m.r.). The number of possible orentiations (spin states) that the nuclear magnet may assume under the influence of the applied magnetic field is determined by the **spin quantum number** of the nucleus. The number of nuclei occupying a particular spin state is known as the population. *See also* **Population inversion** and **Nuclear spin temperature**.

Spin temperature (n.m.r.). *See* **Nuclear spin temperature**.

Sputter-ion pump (Electric discharge getter pump; Getter-ion pump) (m.s., pe.). A special device developed to produce high vacua by the ionization of any residual gases with the simultaneous reaction of ions, atoms and molecules with titanium vapour.[258] After initial reduction of pressure to 10^{-2}-10^{-3} torr with a forepump it is possible to get down to 10^{-10} torr with the sputter-ion pump. The titanium is used in the form of a wire which is gradually unwound[259] as it is vaporized at 2000 °C and gas molecules are effectively coated with the sputtered material. It has the advantage that it requires no cold trap and can be placed close to the vacuum area and is particularly useful for producing the high vacuum conditions required for X-ray photoelectron spectroscopy.

Standard addition method (a.s.). An approach to quantitative analysis which is of wide application in spectroscopy and is particularly applicable to flame photometric determinations. The procedure involves adding a constant amount of the unknown concentration to a series of standards of the pure material and making the mixtures of unknown plus standard to equal volumes. Measurement of the spectra on the flame photometer or spectrometer at a selected wavelength, using each solution individually, produces a series of values which can be plotted against the concentration of the standards. Extrapolation of the line through these points gives an intersection on the concentration line corresponding to the negative value of the unknown concentration. This can also be calculated from the slope of the line without resorting to extrapolation.[260]

The method is used when other substances are present that are likely to introduce errors and interferences. Its value is limited to linear emission/concentration ranges.

Stark effect. When an electric field is applied either parallel or perpendicular to the direction of microwave radiation across a sample the rotational absorption bands of the spectrum are found to be shifted. This is known as the Stark effect. It is due to disturbance of the rotational energy levels of the molecules due to the external field affecting the electric dipole moment. Measurement of the extent of the Stark shift can be used as a basis for the accurate determination of dipole moments.

Stark–Einstein law. Each molecule taking part in a photochemical reaction absorbs one quantum of the radiation leading to the reaction taking place. *See* **Einstein**.

Stefan–Boltzmann constant (σ). The value of this constant applicable in the **Stefan–Boltzmann law** is given by

$$\sigma = \frac{2\pi^5 k^4}{15 h^3 c^2} = 5.670\,32 \times 10^{-8} \text{ W m}^{-2} \text{ K}^{-4}$$

Stefan–Boltzmann law. The total radiation M from unit area of a black body (*see* **Black-body radiation**) in unit time is proportional to the fourth power of the absolute temperature of the body.

$$M = \sigma T^4$$

This relationship is obtained by integrating the equation from the **Planck law of radiation**. *See also* **Wien laws** and **Rayleigh and Jeans law**.

Step sector (em.). Rotation of a step sector placed in the light path of an emission spectrograph produces spectral lines of graduated intensity along the length of each line. This is because the sector reduces the effective light intensity reaching the photographic plate to a fraction equal to the ratio of the area of the cut-away sectors. The device is used to determine the exposure/line density characteristics for the photographic emulsion[261] (*see* Figure 48, page 174; *see also* **Log sector**).

Stigmatic. In the spectroscopic sense this means that light from both horizontal and vertical lines is focused at the same distance so that **astigmatic** distortions do not occur. It is used with reference to diffraction

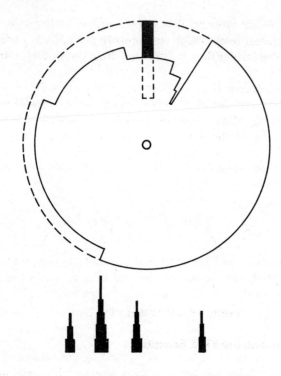

Figure 48. Step sector (appearance of spectrum lines of different intensity indicated below)

gratings and the associated optical systems.

Stokes law (pl.). This law states that the wavelength of a fluorescent emission is always longer than that of the exciting wavelength. This is why substances which appear colourless when viewed in visible light frequently appear coloured when fluorescing under ultraviolet light.

Stokes lines (r.s.). The Raman lines occurring on the longer-wavelength side of the monochromatic radiation source line. They arise from those

Raman transitions in which the final vibrational level is higher than the initial vibrational level. These are normally measured in preference to the **anti-Stokes lines** as they are stronger. *See also* **Raman spectroscopy**.

Stretching vibrations (i.r.). A molecule composed of n atoms exhibits $3n$-6 vibrational modes, or $3n$-5 if it is linear. Of these $n - 1$ modes are valence (or stretching) vibrations. That is, the movement occurs along the bonds between the atoms concerned. The two possible stretching vibrations for a triatomic molecule are shown in the figure below. *See also* **Bending vibrations**.

Figure 49. Stretching vibrations

Surface ionizations. *See* **Field desorption**.

Sweep coils (n.m.r.). Scanning of the spectrum in n.m.r. spectroscopy in older instruments has traditionally been carried out by varying the magnetic field around the sample rather than by altering the radio-frequency field. The sweep over the required field is achieved by passing a current through special sweep coils either wound round the pole pieces or situated between the pole faces and the sample. In the second case these constitute a Helmholtz pair in which the distance between the coils is equal to the radius of the coils, as this produces a good uniform field.

N.B. These are not the **shim coils**.

Symmetry (i.r.). *See* **Axis of symmetry**; **Centre of symmetry**; **Plane of symmetry**.

T

Tantalum boat (a.s.). A microsampling technique developed for atomic absorption spectroscopy in which the sample is placed in a small tantalum dish and inserted into the flame of the spectrometer. This procedure is similar to the **Delves cup** method that was developed particularly for the study of blood samples.

Target (pe.). A specially coated **anode** used to produce **X-rays** as a result of electron bombardment.

Tau values (τ) (n.m.r.). *See* **Parts per million**.

Tesla (T). *See* **Magnetic flux density**.

Tetramethylsilane (n.m.r.). *See* **TMS**.

Thermal emission ion source (m.s.). The evaporation of a substance from a heated surface leads to the formation of positive ions, and this process has been developed as a method for the production of ions for mass spectrometry. The method is used particularly to study isotopic abundance of heavier elements. Operating temperatures as high as 3000 °C are employed and the substance is heated on tungsten or tantalum supports.[262] Useful spectra have been obtained with less than 15 μg of material.

Thermionic emission. The process of producing free electrons by heating metals to high temperatures such that increasing numbers of electrons possess sufficient thermal energy to escape from the surface. Tungsten is the most commonly used metal for **cathodes** for this purpose and is heated to above 2500 K, thoria coated filaments serve the same purpose and operate at temperatures below 2000 K. Electron beams produced by thermionic emission are of importance both in Auger electron spectroscopy (*see* **Auger effect**) and in the production of **X-rays**.

Thermistor (i.r.). A contraction for 'thermal resistor' — substances which change their resistance with temperature. Those developed as infrared detectors are usually known as **bolometers**.

Thermocouple (i.r.). In the thermocouple detector the infrared radiation falls onto a metal junction formed from two dissimilar metals. The change in temperature of the junction causes a current to flow through the metals and this is amplified and fed to a recorder. For the purposes of infrared detection in spectrometers the thermocouple junction is usually welded to a small strip (2 × 0.2 mm^2) of blackened gold leaf.[263] The thermocouple may also be combined with a large number of identical junctions to form a **thermopile**.

Thermopile (i.r.). A thermopile is formed from a number of **thermocouples** joined to produce a series of alternate hot and cold junctions arranged in rows close to each other.[264] The radiation falls on the blackened metal foil forming the heat-sensitive junctions and the temperature difference between this and the combined cold junctions leads to the current flow associated with any thermocouple system. A complete thermopile for use as an infrared detector may have up to twenty hot junctions made from silver and bismuth wires, the foil radiation receiver being blackened gold or platinum.

Tickling (n.m.r.). One of the modified forms of the double-resonance technique in which the extent of **spin–spin coupling** is reduced by bringing about partial perturbation of one of the coupled nuclei. In this process one transition is selectively irradiated, but only to the extent that a complete collapse of the resonance peak does not occur. This tickling procedure enables the coupled nuclei to be identified from the small change in the spectrum without recourse to the full **multiple-resonance** technique.

Time constant (i.r.). Ths is sometimes used as a measure of the slowness of response of a detector in a spectrophotometer. It is normally understood to mean the time needed for the detector to show a drop of 63% when the irradiation is discontinued. *See also* **Response time**.

Time-of-flight mass spectrometer (m.s.). In this instrument ions of different mass/charge ratios are separated according to the time taken to travel along a fixed horizontal path, known as the drift tube (*see* **Drift region**).[265] The ions arrive at the end of the tube in increasing order of their mass/charge ratios. To avoid overlapping spectra the ion source operates in a series of pulses and produces a complete mass spectrum every 100 μs, although the detector is only scanning a small part of the spectrum at any particular period of time (*see* Figure 50, page 179).[266]

Separation of ions is governed by the equation

$$\frac{m}{e} = \frac{2\mathrm{V}t^2}{l^2}$$

from which is can be shown that for the ions

$$\text{velocity} \propto \frac{1}{\sqrt{m}}$$

See also **Continuous mode** and **Pulsed mode**.

TMS (n.m.r.). Tetramethylsilane $(CH_3)_4$ Si is now the accepted standard as the reference signal for proton n.m.r. spectra.[267] All the 12 protons absorb at the same magnetic field strength which is at higher field strength than for the protons in almost every other substance. All **chemical shifts** are then referred to the TMS peak which is specified as absorbing at 0 Hz (0 p.p.m.$\delta \equiv$ 10 p.p.m.τ).

It is a chemically inert liquid boiling at 27 °C. This low boiling point presents difficulties when n.m.r. spectra are determined at elevated temperatures and a secondary standard such as chloroform may be used in place of TMS. As it is also insoluble in heavy water (D_2O) it is necessary to use the water soluble standard **DSS** in this case.

Toronto arc (r.s.). Before the advent of **laser** the Toronto mercury arc source was for many years the standard device for producing monochromatic visible light for **Raman spectroscopy**. Both the 4358 Å and 5461 Å lines were selected for these studies. It has now been almost

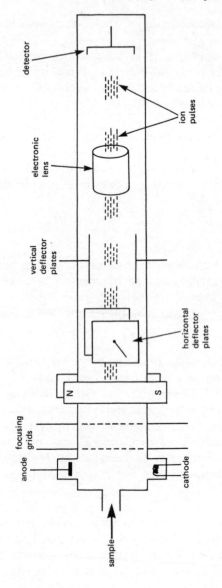

Figure 50. Time-of-flight mass spectrometer

completely displaced because of the stronger sources available from gas lasers.

One of the disadvantages in using arc sources was that filters of aqueous solutions of sodium nitrate or potassium chromate had to be used to prevent interferences due to fluorescence and shorter-wavelength radiations, and as heat filters.

Torr (m.s.). *See* **Pressure, units.**

Transmittance (T) (i.r., u.v.). The ratio of the radiant energy transmitted by a substance to the radiant energy incident upon that substance.

$$T = \frac{I}{I_0}$$

Values obtained for transmittance are usually given as percentages.

The words transmittancy and transmission have been used for this ratio, but transmittance has met with general acceptance and is now recommended.[268] *See also* **Absorbance** and **Molar absorption coefficient.**

Transverse relaxation (T_2) (n.m.r.). One of the two methods by which nuclear spin energy may be lost from one nucleus is by the transfer to another nucleus of the same isotope. This is transverse relaxation and leads to broadening of an n.m.r. absorption peak. It does not result in any change in the numbers of nuclei occupying the different energy levels. The linewidth is inversely proportional to T_2 and is related to the degree of uncertainty in the energy states. The results of transverse relaxation are commonly seen in the n.m.r. spectra of viscous liquids and of solids.

Triplet state (e.s.r., pl.). The condition possible when one of two electrons occupying the same orbit is promoted to an upper orbit. If the spin of the promoted electron is the same as that of the unexcited electron, they are said to be parallel and the spin quantum number (S) is ½ + ½ = 1. This leads to a multiplicity of possible states given by $2S + 1 = 3$ in which the three values correspond to three slightly different energy levels so that the electrons are said to be in a triplet state.

The triplet state is a lower energy state than is the corresponding singlet state and is sometimes referred to as a metastable state. Electrons can remain in the triplet state for extended periods of time, several seconds, and may revert to the ground state by emission of radiation as **phosphorescence.**

Tungsten filament lamp (i.r., u.v.). Although the tungsten filament lamp is usually employed as the source of radiant energy for the visible region (400–800 nm) of the spectrum it may also be used as a source for the near-infrared region as its effective operating range extends as high as 3 μm.

The lamp consists of an incandescent tungsten filament in a glass enclosure. The operating temperature is about 2300 °C and requires a voltage fluctuation of less than 0.01 V to give a constant radiation output.

Incorporation of a small amount of iodine in the tungsten lamp envelope enables operation at higher temperature and an extension of the useful range to below 400 nm.

Twisting vibration (i.r.). **A bending vibration** in which atoms of the same group move in the same plane but in opposite directions to each other. *See also* **Stretching vibrations.**

U

Ultraviolet photoelectron spectroscopy (UPS). *See* **Photoelectron spectroscopy.**

Ultraviolet/visible spectrum. The ultraviolet region of the **electromagnetic spectrum** is usually considered as that section between 10 and 380 nm, whilst the visible region is between 380 and 780 nm. **Spectrophotometers** are manufactured to study both the near-ultraviolet and the visible regions with a continuous scan from 200 to 780 nm. The region with wavelengths shorter than 210 nm is known as the vacuum-ultraviolet region and requires specially modified instruments for study to be made.[269] *See also* **Gratings**; **Prisms**; **Spectrophotometers.**

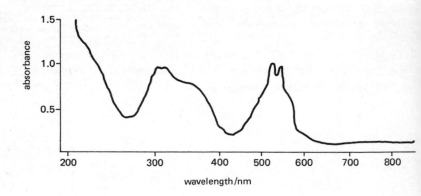

Figure 51. Ultraviolet/visible absorption spectrum for potassium permanganate

U mode (n.m.r.). The absorption signal plotted for the nuclear magnetic resonance spectrum is accompanied by another signal called the 'dispersion' or 'U-mode' signal. This is out of phase with the absorption and is eliminated by the detector phase control. The theoretical Lorentzian line shape (*see* **Lorentzian curve**) of the dispersion curve can be calculated from the **Bloch equations**.

V

Valence vibrations (i.r.). *See* **Stretching vibrations**.

Variable path-length cell (i.r.). *See* **Liquid cell**.

Velocity of light (c). This is identical for all wavelengths and is usually given as the velocity of light in a vacuum with the value

$$c = 2.997\,9245\,8 \times 10^8 \text{ m s}^{-1} \approx 3 \times 10^{10} \text{ cm s}^{-1}$$

Vibrational spectra (i.r., r.s.). Vibrational spectra are observed when the absorption of radiant energy causes a change in the energy of a molecular vibration. These transitions are accompanied by changes in the rotational state of the molecule and are observed in the region 4000–400 cm^{-1} (2.5–25 μm) as **infrared spectra** and **Raman spectra**.

Virtual coupling (n.m.r.). If a nucleus is coupled to another nucleus that in turn is coupled to a third nucleus possessing a similar **chemical shift** as the second nucleus, a long-range coupling effect between the first and third nuclei occurs. As a result the simple **multiplet** pattern for the absorption of the first proton is complicated by further splitting which produces a distorted, non-uniform appearance to the observed absorption spectrum. Virtual coupling of this type is common to molecules possessing long alkyl chains.

Visible spectrum. *See* **Electromagnetic spectrum** and **Ultraviolet/ visible spectrum**.

v mode (n.m.r.). The normal abosprtion signal measured and recorded in n.m.r. spectroscopy. It is out of phase with the dispersion, **u-mode**, signal.

W

Wadsworth mounting (em.). Of the different forms of mounting employed for emission spectrographs the Wadsworth has been particularly successful as it is **stigmatic** while most other emission mountings are **astigmatic**. It achives this by employing a concave diffraction grating in place of the conventional plane surface **grating**. The arrangement is illustrated in Figure 52, page 186.

Wagging vibration (i.r.). Of the four types of **bending vibrations** the wagging vibration is one in which there is no change in the angle between the bonds of the vibrating group. This is because the vibration takes place in a plane at right angles to the plane of the group.

Watson–Biemann separator (m.s.). This particular type of **GC/MS interface** consists of a long (20 cm), narrow (8 mm) porous glass diffusion tube. Effluent gases from the gas chromatograph pass along the tube and the carrier gas diffuses through the pores more rapidly than do the heavier organic molecules.[270] As a result a fifty-fold enrichment is obtained with almost 50 per cent recovery of the sample.

Wave equation. *See* **Schrödinger wave equation.**

Waveguides (e.s.r.). Rectangular (or sometimes circular) cross section metal tubes used to direct the microwaves generated by the **klystron oscillator** used in an e.s.r. spectrometer. They are made from brass plated with silver.

The waves pass along the waveguides by reflection from the walls of the tube. Waveguides are employed because at microwave frequencies transmission losses are substantial if conventional means are used. The dimensions of the waveguides influence the field patterns created by the microwaves and it is normal to employ a waveguide with a cross-section dimension comparable with the frequency being generated by the klystron (e.g. approximately 2 cm × 1 cm).

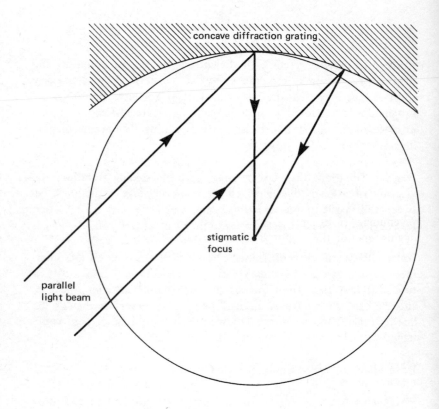

Figure 52. Wadsworth mounting

Wavelength. The linear distance measured along the line of propogation of the wave between any two points that are in phase on adjacent waves. For the sake of convenience the measurement is normally taken between maxima or minima on the wave diagram. Electromagnetic radiation is a sinusoidal form of the type shown in Figure 53 (*a*) but other wave forms such as square waves (*b*) and half waves (*c*) can be created by electronic means.

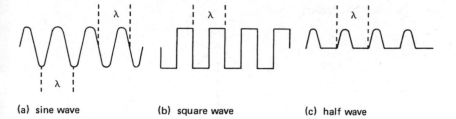

(a) sine wave   (b) square wave   (c) half wave

Figure 53. Waves and wavelength

Wavenumber ($\tilde{\nu}$ or σ) (i.r.). In addition to the above two symbols used for this function the symbol k is employed in solid-state physics.

It is related to **wavelength** and **frequency** by the following equation:

$$\tilde{\nu}\,(\sigma) = \frac{\nu}{c} = \frac{1}{\lambda}$$

and is usually expressed in reciprocal centimetres (cm^{-1}).

Weber (Wb). The **SI unit** of magnetic flux (Φ). It is related to other units as follows:

$$Wb = m^2 \ kg \ s^2 \ A^{-1} = V \ s$$

and 1 Wb = 10^8 maxwell.

Wedge. *See* **Attenuator**.

Wheatstone bridge circuit. In the circuit shown below if the four resistances satisfy the equation $R_1/R_2 = R_3/R_4$ then no current flows through the galvanometer. This type of circuit is the basis of potentiometry and is used in **spectrophotometers** incorporating resistance type detectors such as the **bolometer**. The change in resistance of the detector throws the bridge circuit out of balance and the resulting signal can be fed to a recorder.

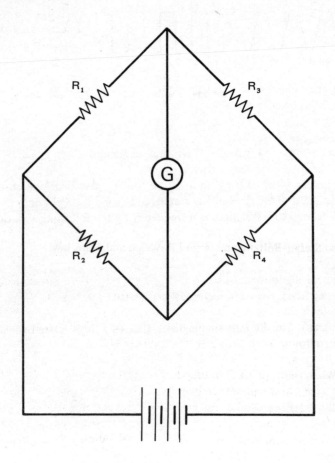

Figure 54. Wheatstone bridge circuit

Wien laws. These three laws on **black-body radiation** pre-date the **Planck law of radiation** but can be derived from it. They are as follows.

(*a*) The wavelength of the maximum of the intensity of radiation from the black-body source is inversely proportional to the absolute temperature.

$$\lambda_{max} T = \text{constant}$$

(*b*) The radiant exitance at the wavelength maximum is proportional to the fifth power of the absolute temperature.

$$M_{\lambda max} T^{-5} = \text{constant}$$

(*c*) The spectral radiant exitance is given by

$$M_\lambda = \frac{2\pi hc^2 \lambda^{-5}}{e^{hc/kT\lambda} - 1}$$

See also **Stefan–Boltzmann law** and **Rayleigh and Jeans law.**

Wiggle-beat method (n.m.r.). Rapid scanning of two closely spaced absorption signals leads to superimposition of the **ringing** of the two signals to form a beat pattern. The frequency of the beat is equal to the separation of the signals and arises from the inability of the nuclear magnetization to keep pace with the applied field.

Woodward rules (u.v.). Structural modifications to a chromophoric system (*see* **Chromophore**) frequently show a simple additive relationship. By assigning a value (in nm) to each functional part of a molecule and adding them together it is possible to predict where the major absorption for the system will arise.[271] Woodward[272] applied this procedure extensively to his study of sterols and related compounds.

Work function (ϕ) (pe.). The minimum (or threshold) photon energy $h\nu_0$ necessary for photoionization is known as the work function in solids and corresponds to the **ionization energy** for gaseous substances.

The work function is approximately equal to the **Fermi energy** and corresponds to the energy required to remove an electron from the Fermi level to a distance greatly in excess of the interatomic distance but small relative to the crystal size.[273]

Wratten filter. A filter used in simple colorimeters which consists of a thin film of dyed gelatine sandwiched between two pieces of glass. The filter deteriorates with time due to decomposition of the dyes.

X

Xenon arc lamp (a.s., u.v.). A radiation source employed in u.v./visible and in fluorescent spectroscopy when a very high level of continuous illumunation is required. A 150 W xenon arc source operating on d.c. is most commonly used, the useful radiation range being between 220 and 850 nm. It has been used to obtain rapid analyses by atomic fluorescence[274] with detection limits below 5 p.p.m. for many elements.[275]

X-ray fluorescence (XRF) (pe.). When an electron 'hole' is produced in a core sub-shell it can decay by three mechanisms. In the XRF decay process the vacancy is filled by movement of an electron from another sub-shell possessing a lower binding energy, at the same time energy is released as a photon which is in the X-ray region of the electromagnetic spectrum (Figure 55, page 192). X-ray fluorescence is the basis of X-ray emission spectroscopy in which the spectra of the emitted X-rays is studied. *See also* **Auger effect** and **Coster-Kronig process**.

X-rays (pe.). The X-ray region of the electromagnetic spectrum is between 6×10^{19} Hz $- 1 \times 10^{17}$ Hz (5×10^{-12} m $- 3 \times 10^{-9}$ m), X-rays being produced when streams of electrons from tungsten or thoria **cathodes** strike metallic targets. Most X-rays used for **photoelectron spectroscopy** are of low photon energy (around 1000 eV) and are called soft-X-rays. Hard-X-rays, produced from metals such as copper and chromium, have high photon energies of 5000–8000 eV. Photoelectron spectroscopy is usually carried out using the Kα lines emitted by aluminium or magnesium anodes; the designation Kα shows that the emission of the X-rays occurs as a result of a K hole (in a 1s shell) being filled by an electron moving from a 2p shell. Similarly, what are known as L lines arise from vacancies in L-orbitals being filled by outer electrons dropping back. For use in photoelectron spectroscopy the broad band **Bremsstrahlung** is removed by passing through a quartz **monochromator**.

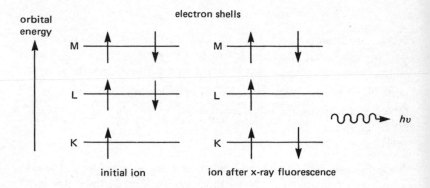

Figure 55. X-ray fluorescence process

X-ray photoelectron spectroscopy (XPS). *See* **Photoelectron spectroscopy**.

X, Y **and** *Z* **gradients** (n.m.r.). The three mutually perpendicular axes are used to relate the parts of the **probe** in the n.m.r. spectrometer to the magnet system. The X axis is taken as that perpendicular to the sample tube and parallel to the magnet faces (pole pieces); the Y axis is along the length of the sample tube; the Z axis is through the centres of the pole pieces and perpendicular to the sample tube.

Z

Zeeman effect. The effect of overcoming the normal degenercy of electron spin states by applying a magnetic field which can interact with the magnetic moment of the electron. The Zeeman effect is observed when atoms are subjected to a powerful magnetic field resulting in the spectral lines being split into a number of component lines.

The energy level difference is given by

$$\Delta E = m_1 g \mu_b B$$

The energy levels are sometimes referred to as Zeeman levels and in effect it is transitions between these levels that are measured in **electron spin resonance** spectroscopy.

References

1. 'Quantities, Units and Symbols', *A Report by the Symbols Committee of the Royal Society*, London (1975). *Addenda* (1981)
2. H.M.N. Irving, H. Freiser and T.S. West (eds), *Compendium of Analytical Nomenclature, Definitive Rules 1977*, IUPAC, Pergamon Press, Oxford and New York (1978)
3. D.H. Whiffen (ed.), *Manual of Symbols and Terminology for Physicochemical Quantities and Units*, IUPAC, Pergamon Press, Oxford and New York (1979)
4. C. Moser and A.I. Kohlenberg, *J. Chem. Soc.*, 804 (1951)
5. 'Spectrometry Nomenclature', *Anal. Chem.*, **37**, 1814 (1965)
6. J.H. Beynon and A.E. Williams, *Mass and Abundance Tables for Use in Mass Spectrometry*, Elsevier, London (1963)
7. A. Barrie, 'Instrumentation for Electron Spectroscopy', *Handbook of X-ray and Ultraviolet Photoelectron Spectroscopy* (ed. D. Briggs), Heyden, London, p. 79 (1978)
8. C.A. Parker, *Proc. Soc. Anal. Chem. Conf., Nottingham* (ed. P.W. Shallis), W. Heffer and Sons Ltd, London (1955)
9. H.C. Kahn, *J. Chem. Educ.*, **43**, A7, A103 (1966)
10. T.S. West, *Chem. Ind.*, 387 (1970)
11. W.J. Price, *Int. Lab.*, July/Aug., 89 (1980)
12. W.J. Price, *Spectrochemical Analysis by Atomic Absorption*, Heyden, London and New York (1980)
13. R.M. Dagnall, K.C. Thompson and T.S. West, *Anal. Chim. Acta*, **36**, 269 (1966)
14. S.J. Weeks, H. Haraguchi and J.D. Winefordner, *Anal. Chem.*, **50**, 360 (1978)
15. J.J. Balmer, *Ann. Physik*, **25**, 80 (1885)

16. J. Fahrenfort, *Spectochim. Acta,* **17**, 698 (1961)
17. N.J. Harrick, *Internal Reflection Spectroscopy*, Interscience, New York (1967)
18. P.A. Wilks, *Int. Lab.*, July/Aug., 47 (1980)
19. J.P. Barthelemy, A. Copin, R. Deleu and J.L. Closset, *Talanta,* **26**, 885 (1979)
20. P. Auger, *J. Phys. Radium,* **6**, 205 (1925)
21. T.A. Carlson, *Photoelectron and Auger Spectroscopy*, Plenum Press, New York (1976)
22. S. Nagakura and J. Tanaka, *J. Chem. Phys.,* **22**, 236 (1954)
23. R.M. Silverstein, G.C. Bassler and T.C. Morrill, *Spectrometric Identification of Organic Compounds*, 3rd edn, Wiley, New York (1974)
24. W.F. Forbes and R. Shilton, in *Symp. Spectrosc., ASTM Spec. Tech. Publ.*, no. 269, 176 (1960)
25. A. Burawoy, *J. Chem. Soc.*, 1177 (1939)
26. K. Bowden and E.A. Braude, *J. Chem. Soc.*, 1068 (1952)
27. L. Doub and J.H. Vandenbelt, *J. Am. Chem. Soc.,* **69**, 2714 (1947)
28. H.B. Klevens and J.R. Platt, *J. Am. Chem. Soc.,* **71**, 1714 (1949)
29. C. Moser and A.I. Kohlenberg, *J. Chem. Soc.*, 804 (1951)
30. E. Clar, *Aromatische Kohlenwasserstoffe*, Springer-Verlag, Berlin (1952)
31. N. Wright, *Anal. Chem.,* **13**, 1 (1941)
32. G. Pirlot, *Bull. Soc. Chim. Belg.,* **58**, 28, 48 (1949)
33. C. Moser and A.I. Kohlenberg, *J. Chem. Soc.*, 804 (1951)
34. K. Bowden and E.A. Braude, *J. Chem. Soc.*, 1068 (1952)
35. J.H. Lambert, *Photometria sive de Mensura et Gradibus Luminis, Colorum et Umbrae*, Angsburg (1760); repr. in Otswald, *Klassiker der Exakten Wissenschaften*, no. 32, 64 (1892)
36. M. Bouguer, *Essai d'Optique sur la Gradation de la Lumière*, Paris (1729); see also Otswald, *Klassiker der Exakten Wissenschaften*, no. 33, 58 (1892); M. Bouguer, *Traité d'Optique sur la Gradation de la Lumière, Ouvrage Posthume*, Pub. de Lacaille (1760)
37. A. Beer, *Ann. Physik Chem.* (J.C. Poggendorf), **86**, 78 (1852)
38. F. Bernard, *Ann. Chim. Phys.,* **35**, [3], 385 (1852)

39.   D.R. Malinin and J.H. Yoe, *J. Chem. Educ.*, **38**, 129 (1961)
40.   F.H. Lohman, *J. Chem. Educ.*, **32**, 155 (1955)
41.   F. Bloch, *Phys. Rev.*, **70**, 460 (1946)
42.   F. Bloch, *Phys. Rev.*, **102**, 104 (1956)
43.   R.M. Milton, *Chem. Rev.*, **39**, 419 (1946)
44.   W.H. Brattain and J.A. Becker, *J. Opt. Soc. Am.*, **36**, 354 (1946)
45.   N. Fuson, *J. Opt. Soc. Am.*, **38**, 845 (1948)
46.   B.D. Flockhart and R.C. Pink, *Talanta*, **12**, 546 (1965)
47.   G. Powles, A. Hartland and J.A.E. Kail, *J. Polymer Sci.*, **55**, 361 (1961)
48.   R.K. Harris and K.J. Packer, *European Spectroscopy News*, [21], 37 (1978)
49.   *Tables of Wavenumbers for the Calibration of Infrared Spectrometers* IUPAC, Butterworths, London (1961)
50.   *Pure Appl. Chem.*, **1**, [4], 537–699 (1961)
51.   R.G. Anderson, I.S. Maines and T.W. West, *Anal. Chim. Acta*, **51**, 335 (1970)
52.   J.F. Alder and T.S. West, *Anal. Chim. Acta*, **51**, 365 (1970)
53.   C. Moser and A.I. Kohlenberg, *J. Chem. Soc.*, 804 (1951)
54.   C.A. Parker, *Proc. Roy. Soc. (London), Ser. A*, **220**, 104 (1953)
55.   J.N. Pitts, G.W. Cowell and D.R. Burley, *Environmental Sci. Tech.*, **2**, 435 (1968)
56.   L.W. Sieck, *Chem. Brit.*, **16**, 38 (1980)
57.   R.M. Silverstein, G.C. Bassler and T.C. Morrill, *Spectrometric Identification of Organic Compounds*, 3rd edn, Wiley, New York (1974)
58.   T.A. Carlson, *Photoelectron and Auger Spectroscopy*, Plenum Press, New York (1976)
59.   C.C. Hinckley, *J. Am. Chem. Soc.*, **91**, 5160 (1969)
60.   A.F. Cockerill, G.L.O. Davies and D.M. Rackham, *Chem. Rev.*, **73**, 553 (1973)
61.   W.D. Horrocks and J.P. Sipe, *J. Am. Chem. Soc.*, **93**, 6800 (1971)
62.   J. Briggs, F.A. Hart, G.P. Moss and E.W. Randall, *Chem. Commun.*, 364 (1971)
63.   A.A. Mills, *Chem. Brit.*, **16**, 69 (1980)

64. J.A. Schellman, *Chem. Rev.*, **75**, 323 (1975)
65. T.A. Crone and N. Purdie, *Anal. Chem.*, **53**, 17 (1981)
66. M. Czerny and A.F. Turner, *Z. Physik*, **61**, 792 (1930)
67. W.G. Fastie, *J. Opt. Soc. Am.*, **42**, 641 (1952)
68. H. Ebert, *Wied Ann.*, 489 (1889)
69. C. Moser and A.I. Kohlenberg, *J. Chem. Soc.*, 804 (1951)
70. H.T. Delves, *Analyst*, **95**, 431 (1970)
71. H.T. Delves, *Atomic Absorption Newsletter*, **12**, 50 (1973)
72. E.L. Henn, *Atomic Absorption Newsletter*, **12**, 109 (1973)
73. J.E. Cahill, *Int. Lab.*, Jan./Feb., 64 (1980)
74. W.A. Wolstenholme and R.M. Elliott, *Int. Lab.*, May/June, 55 (1976)
75. K. Bowden and E.A. Braude, *J. Chem. Soc.*, 1068 (1952)
76. R.M. Dagnall, K.C. Thompson and T.S. West, *Talanta*, **14**, 551 (1967)
77. K.M. Aldous, D. Alder, R.M. Dagnall and T.S. West, *Lab. Pract.*, **19**, 589 (1970)
78. K. Morita, H. Sunahara, T. Ishizuka, N. Nakayama and S. Yamada, *J. Spectrosc. Soc. Japan*, **16**, 169 (1968); *Chem Abstracts*, **69**, 56728 (1968)
79. B. Findeisen and W. Schuffenhauser, *Neue Huette*, **15**, 437 (1970): *Chem. Abstracts*, **73**, 116009 (1970)
80. G.W. Ewing, *J. Chem. Educ.*, **46**, 2, A69 (1969)
81. H. Ehrhardt, L. Langhans, F. Londer and H.S. Taylor, *Phys. Rev.*, **A173**, 222 (1968)
82. C.E. Kuyatt and J.A. Simpson, *Rev. Sci. Instrum.*, **38**, 102 (1967)
83. J.D. Lee, *Rev. Sci. Instrum.*, **44**, 893 (1973)
84. J.W. McGowan, D.A. Vroom and A.R. Comeaux, *J. Chem. Phys.*, **51**, 5626 (1969)
85. C.E. Kuyatt and J.A. Simpson, *Rev. Sci. Instrum.*, **38**, 103 (1967)
86. A. Barrie, *J. Electron Spectroscopy*, **7**, 75 (1975)
87. J.E. Wertz and J.R. Bolton, *Electron Spin Resonance, Elementary Theory and Applications*, McGraw-Hill, New York (1972)
88. M. Symons, *Chemical and Biochemical Aspects of Electron Spin Resonance Spectroscopy*, Van Nostrand Reinhold, New York (1978)

89. J.M. Ottaway and F. Shaw, *Analyst,* **101**, 582 (1976)
90. R.E. Sturgeon, *Anal. Chem.*, [14], **49**, 1255A (1977)
91. S. Greenfield and H.M. McGeachin, *Chem. Brit.*, **16**, 653 (1980)
92. G. Feher, *Phys. Rev.*, **103**, 834 (1956)
93. M.M. Dorio and J.H. Freed (eds), *Multiple Electron Resonance Spectroscopy*, Plenum Press, New York (1979)
94. K. Siegbahn *et al.*, *Nova Acta Regiae Sci. Ups. Ser.*, **20**, iv (1967)
95. D. Briggs, in *Electron Spectroscopy Theory, Techniques and Applications* (eds C.R. Brundle and A.D. Baker), Academic Press, New York, vol. 3 (1979)
96. B. Stevens and E. Hutton, *Spectrochim. Acta,* **18**, 425 (1962)
97. B. Stevens, M.S. Walker and E. Hutton, *Proc. Chem. Soc.*, 62 (1963)
98. M. Faraday, *Phil. Mag.*, **28**, 294 (1846)
99. A.C. Hardy and F.H. Perrin, *The Principles of Optics*, McGraw-Hill, New York, p. 556 (1932)
100. Féry, *J. Phys.*, **9**, 762 (1910); *Astrophys. J.*, **34**, 79 (1911)
101. A.G. Brenton and J.H. Beynon, *European Spectroscopy News*, **29**, 39 (1980)
102. R.P. Nyquist and R.O. Kagel, *Infrared Spectra of Inorganic Compounds*, Academic Press, New York (1971)
103. H. Budzikiewicz, C. Djerassi and D.H. Williams, *Interpretation of Mass Spectra of Organic Compounds*, Holden-Day, San Francisco (1964)
104. J. Bassett, R.C. Denney, G.H. Jeffrey and J. Mendham, *Vogel's Textbook of Quantitative Inorganic Analysis*, 4th edn, Longmans, London and New York, pp. 710, 826 (1978)
105. E.E. Pickett and S.R. Koirtyohann, *Anal. Chem.*, **41**, [14], 28A (1969)
106. E. Mavrodineanu, *Spectrochim. Acta,* **17**, 1016 (1961)
107. K. Szivós, E. Pungor and L. Kiss, *Talanta*, **26**, 849 (1979)
108. L.R.P. Butler and A. Fulton, *Appl. Opt.*, **7**, 2131 (1968)
109. B.L. Van Duuren and T.L. Chan, *Advan. Anal. Chem. Instrum.*, **9**, 387 (1971)
110. W.H. Melhuish, *J. Opt. Soc. Am.*, **52**, 1256 (1962)
111. C.A. Parker, *Photoluminescence of Solutions*, Elsevier, London,

p. 128 (1969)

112. W. Gordy, *J. Chem. Phys.*, **14**, 305 (1946)

113. R.R. Ernst and W.A. Anderson, *Rev. Sci. Instrum.*, **37**, 93 (1966)

114. P.L. Richards, *J. Opt. Soc. Am.*, **54**, 1474 (1964)

115. H. Ishida and J.L. Koenig, *Int. Lab.*, July/Aug., 89 (1980)

116. W.J. Hurley, *J. Chem. Educ.*, **43**, 236 (1966)

117. K. Krishnan, R.H. Brown, S.L. Hill, S.C. Simonoff, M.L. Olson and D. Kuehl, *Int. Lab.*, May/June, 66 (1981)

118. R.R. Ernst and W.A. Anderson, *Rev. Sci. Instrum.*, **37**, 93 (1966)

119. J. Franck, *Trans. Faraday Soc.*, **21**, 536 (1926)

120. A.G. Gaydon and H.G. Wolfhard, *Flames, Their Structure, Radiation and Temperature*, Chapman and Hall, London (1953)

121. E. Mavrodineanu, *Spectrochim. Acta*, **17**, 1016 (1961)

122. M.J. O'Neal and T.P. Wier, *Anal. Chem.*, **23**, 830 (1951)

123. D. Welti, *Infrared Vapour Spectra*, Heyden, New York (1970)

124. M.D. Erickson, *Applied Spectroscopy Revs.*, **15**, 261 (1979)

125. D.I. Rees, *Talanta*, **16**, 903 (1969)

126. S. Silvermann, *J. Opt. Soc. Am.*, **38**, 989 (1948)

127. B.L. Henke, *Adv. X-ray Anal.*, **4**, 244 (1961)

128. J.S. Cartwright, G. Sebens and W. Slavin, *Atomic Absorption Newsletter*, **5**, 22 (1966)

129. J.V. Sullivan and A. Walsh, *Spectrochim. Acta*, **21**, 721 (1965)

130. W.G. Jones and A. Walsh, *Spectrochim. Acta*, **11**, 249 (1960)

131. A. Paschen, *Physik*, **50**, 901 (1916)

132. W.J. Price, *Analytical Atomic Absorption Spectrometry*, Heyden, London, p. 37 (1972)

133. C.A. Parker, *Proc. Soc. Anal. Chem. Conf., Nottingham* (ed. P.W. Shallis), W. Heffer and Sons Ltd, London (1955)

134. A.D. Cross and R.A. Jones, *An Introduction to Practical Infra-red Spectroscopy*, 3rd edn, Butterworths, London, p. 43 (1969)

135. E.B. Baker, *J. Chem. Phys.*, **37**, 911 (1962); **45**, 609 (1966)

136. R.K. Harris, *Chem. Soc. Rev.*, **5**, 1 (1976)

137. F. Bloch, W.W. Hansen and M. Packard, *Phys. Rev.*, **70**, 474 (1946)

138. R. Mavrodineanu and R.C. Hughes, *Spectrochim. Acta*, **19**, 1309 (1963)

139. C.D. West and D.N. Hume, *Anal. Chem.*, **36**, 412 (1964)
140. G.F. Kirkbright and H.M. Tinsley, *Talanta*, **26**, 41 (1979)
141. S. Greenfield and H.M. McGeachin, *Chem. Brit.*, **16**, 653 (1980)
142. H. Levinstein, *Anal. Chem.*, **41**, 14, 81A (1969)
143. K.D. Möller and W.G. Rothschild, *Far-infrared Spectroscopy*, Wiley, New York (1971)
144. S.E. Wiberley, J.W. Sprague and J.E. Campbell, *Anal. Chem.*, **29**, 210 (1957)
145. R.B. Barnes, R.C. Gore, E.F. Williams, S.G. Linsley and E.M. Peterson, *Anal. Chem.*, **19**, 620 (1947)
146. J.H. Beynon, *Mass Spectrometry and its Application to Organic Chemistry*, Elsevier, London, p. 459 (1960)
147. G.W. Ewing, *J. Chem. Educ.*, **46**, 2, A69 (1969)
148. J.H. Beynon and A.E. Williams, *Mass and Abundance Tables for Use in Mass Spectrometry*, Elsevier, London (1963)
149. H.M.N. Irving, H. Freiser and T.S. West (eds), *Compendium of Analytical Nomenclature, Definitive Rules 1977*, IUPAC, Pergamon Press, Oxford and New York (1978)
150. D.H. Whiffen (ed.), *Manual of Symbols and Terminology for Physicochemical Quantities and Units*, IUPAC, Pergamon Press, Oxford and New York (1979)
151. R. Ryhage, *Anal. Chem.*, **36**, 759 (1964)
152. M. Karplus, *J. Chem. Phys.*, **30**, 11 (1959)
153. M. Karplus, *J. Am. Chem. Soc.*, **85**, 2870 (1963)
154. *Trans. of the Joint Committee for Spectroscopy, J. Opt. Soc. Am.*, **43**, 410 (1953)
155. *Trans. of the Joint Committee for Spectroscopy, J. Opt. Soc. Am.*, **46**, 145 (1956)
156. W.F. Forbes and R. Shilton, in *Symp. Spectrosc., ASTM Spec. Tech. Publ.*, no. 269, 176 (1960)
157. G.W. Ewing, *J. Chem. Educ.*, **46**, 2, A69 (1969)
158. W.A. Chupka and M.G. Inghram, *J. Phys. Chem.*, **59**, 100 (1955)
159. T. Koopmans, *Physical*, **1**, 104 (1934)
160. W.B. Rodney and I.H. Malitson, *J. Opt. Soc. Am.*, **46**, 956 (1956)
161. M. Bouguer, *Essai d'Optique sur la Gradation de la Lumière*, Paris (1729); see also Otswald, *Klassiker der Exakten Wissenschaften*,

no. 33, 58 (1892); M. Bouguer, *Traité d'Optique sur la Gradation de la Lumière, Ouvrage Posthume*, Pub. de Lacaille (1760)

162. J.H. Lambert, *Photometria sive de Mensura et Gradibus Luminus, Colorum et Umbrae*, Angsburg (1760); repr. in Otswald, *Klassiker der Exakten Wissenschaften*, no. 32, 64 (1892)

163. S. Kimel and S. Speiser, *Chem. Rev.*, **77**, 437 (1977)

164. J.E. Thomas, M.J. Kelly, J.-P. Monchalin, N.A. Kurnit and A. Javan, *Rev. Sci. Instrum.*, **51**, 240 (1980)

165. J.K. Burdett and M. Poliakoff, *Chem. Soc. Rev.*, **3**, 293 (1974)

166. V.V. Panteleev and A.A. Yankovskii, *J. Appl. Spectrosc. (Zh. Prikli. Spectrosk.)*, **3**, 260 (1965)

167. D.C. Damoth, *Modern Aspects of Mass Spectrometry* (ed. R.I. Reed), Plenum Press, New York, p. 49 (1968)

168. F.J. Vastola and A.J. Pirone, *Advan. Mass Spectrometry*, **4**, 107 (1968)

169. S.P.S. Porto and D.I. Wood, *J. Opt. Soc. Am.*, **52**, 251 (1962)

170. D. Vidrine and D. Warren, *Fourier Transform Infrared Spectroscopy*, **2**, 129 (1979)

171. W.J. McFadden, *J. Chromatog. Sci.*, **50**, 97 (1980)

172. P.J. Arpino, B.G. Dawkins and F.W. McLafferty, *J. Chromatog. Sci.*, **12**, 574 (1974)

173. W.J. McFadden, H.L. Schwartz and S. Evans, *J. Chromatog.*, **122**, 389 (1975)

174. R.G. Christensen, H.S. Hertz, S. Meilselman and E. White, *Anal. Chem.*, **53**, 171 (1981)

175. J.S. Griffith and L.E. Orgel, *Quart. Rev.*, **11**, 381 (1957)

176. E.L. Grove (ed.), *Analytical Emission Spectroscopy*, vol. 1, Dekker, New York, part 1, p. 32 (1971)

177. L.H. Germer, *Sci. Amer.*, [3], **212**, 32 (1965)

178. J.B. Pendry, *Low Energy Electron Diffraction*, Academic Press, London and New York (1974)

179. J.A. Gilpin and F.W. McLafferty, *Anal. Chem.*, **29**, 990 (1957)

180. D.G.I. Kingston, J.T. Bursey and M.M. Bursey, *Chem. Rev.*, **74**, 215 (1974)

181. F.W. McLafferty, *Interpretation of Mass Spectra*, 3rd edn, University Science Books, Mill Valley, California (1980)

202   References

182. H. Massman, *Spectrochim. Acta,* **23B**, 215 (1968)
183. J.J. Thomson, *Phil. Mag.,* **21**, 225 (1911)
184. F.W. Aston, *Phil. Mag.,* **38**, 707 (1919)
185. J.H. Beynon, *Mass Spectrometry and its Applications to Organic Chemistry,* Elsevier, London, p. 459 (1960)
186. J. Mattauch and R.F.K. Herzog, *Z. Physik,* **89**, 786 (1934)
187. J.R. Partington, *An Advanced Treatise on Physical Chemistry,* vol. 1, Longmans, London, p. 303 (1949)
188. J.H. Beynon, *Advan. Mass Spectrometry,* **4**, 123 (1968)
189. J.H. Beynon, R.A. Saunders and A.E. Williams, *Table of Metastable Transitions for Use In Mass Spectrometry,* Elsevier, London (1965)
190. A.G. Brenton and J.H. Beynon, *European Spectroscopy News,* **29**, 39 (1980)
191. R.N. Jones and A. Nadeau, *Spectrochim. Acta,* **12**, 183 (1958)
192. D.S. Erley, 'Long-path Infrared Microcell', in *Spectroscopic Tricks* (ed. L. May), A. Hilger Ltd, London, p. 185 (1968)
193. D.R. Lide, in *Advan. Anal. Chem. Instrum.* (eds C.N. Reilley and F.W. McLafferty), **5**, 235 (1966)
194. A. Barrie and N.E. Christensen, *Phys. Rev.,* **314**, 2442 (1976)
195. R.L. Mössbauer, *Z. Physik,* **151**, 124 (1958)
196. L. May, *Appl. Spectrosc.,* **23**, 204 (1969)
197. V.I. Goldanskii and R.H. Herber, *Chemical Applications of Mössbauer Spectroscopy,* Academic Press, New York (1968)
198. R.H. Herber, *Sci. Am.,* **225**, 86 (1971)
199. A. Vĕrtes, L. Korecz and K. Burger, *Mössbauer Spectroscopy,* Elsevier, Amsterdam, Oxford and New York (1979)
200. J.D. Baldeschwieler and E.W. Randall, *Chem. Rev.,* **63**, 81 (1963)
201. W. McFarlane, *Chem. Brit.,* **5**, 142 (1969)
202. J. Stupar and J.B. Dawson, *Appl. Opt.,* **7**, 1351 (1968)
203. G. Uny and J. Spitz, *Spectrochim. Acta,* **25B**, 391 (1970); *Chem. Abstracts,* **73**, 132441 (1970)
204. C.W. Munday, *J. Sci. Instrum.,* **25**, 418 (1948)
205. A.O. Nier and E.G. Johnson, *Phys. Rev.,* **91**, 10 (1953)
206. J.R. Majer, *The Mass Spectrometer,* Wykeham Publications (London) Ltd, London (1977)

207. R.C. Denney, *Chem. Brit.*, **16**, 428 (1980)
208. R.M. Lynden-Beil and R.K. Harris, *Nuclear Magnetic Resonance Spectroscopy*, Nelsons, London (1969)
209. M. Kubo and D. Nakamura, *Advan. Inorg. Radiochem.*, **8**, 257 (1966)
210. C.T. O'Konski, 'Nuclear Quadropole Resonance Spectroscopy', in *Determination of Organic Structures by Physical Methods* (eds F.C. Nachod and W.D. Phillips), vol. 2, Academic Press, New York, chap. 11 (1962)
211. G.C. Lyle and R.E. Lyle, 'Optical Rotary Dispersion', in *Determination of Organic Structures by Physical Methods* (eds F.C. Nachod and W.D. Phillips), vol. 2, Academic Press, New York, chap. 1 (1962)
212. R.A. Dwek, R.E. Richards and D. Taylor, *Ann. Rept. Nucl. Magnetic Resonance Spectrosc.*, **2**, 293 (1969)
213. K.H. Hauser and D. Stehlik, *Advan. Magnetic Resonances*, **3**, 79 (1968)
214. J.R. Dyer, *Applications of Absorption Spectroscopy of Organic Compounds*, Prentice-Hill, Englewood Cliffs, NJ, p. 52 (1965)
215. D.S. Parasnis, *Magnetism*, Hutchinson, London, p. 50 (1961)
216. G.M. Barrow, *Physical Chemistry*, McGraw-Hill, New York, p. 307 (1961)
217. A.G. Bell, *Am. J. Sci.*, **20**, 305 (1880)
218. G.F. Kirkbright and S.L. Castledon, *Chem. Brit.*, **16**, 661 (1980)
219. A. Rosencwaig and A. Gersho, *J. Appl. Phys.*, **47**, 64 (1976)
220. L.W. Buggraf and D.E. Leyden, *Anal. Chem.*, **53**, 759 (1981)
221. R.A. Smith, *Advan. Phys.*, **2**, 321 (1953)
222. D.W. Turner, *Chem. Brit.*, **4**, 435 (1968)
223. D. Briggs (ed.), *Handbook of X-ray and Ultraviolet Photoelectron Spectroscopy*, Heyden, London (1978)
224. J.B. Pearce, K.A. Gause, E.F. Mackey, K.K. Kelly, W.G. Fastie and C.A. Barth, *Appl. Opt.*, **10**, 805 (1971)
225. M.J.E. Golay, *Rev. Sci. Instrum.*, **18**, 347, 357 (1947)
226. M.J.E. Golay, *Rev. Sci. Instrum.*, **20**, 816 (1949)
227. G.K.T. Conn and G.K. Eaton, *J. Opt. Soc. Am.*, **44**, 553 (1954)
228. R. Newmann and R.S. Halford, *Rev. Sci. Instrum.*, **19**, 270 (1948)

229. G.R. Bird and M. Parrish, *J. Opt. Soc. Am.,* **50**, 886 (1960)
230. *Perkin Elmer Instrum. News,* **16**, [3] , 5 (1966)
231. J.B. Young, H.H. Graham and E.W. Peterson, *Appl. Opt.,* **4**, 1023 (1965)
232. J.G. Calvert and J.N. Pitts, *Photochemistry*, Wiley, New York, p. 728 (1966)
233. A.R. Fairbairn, *Rev. Sci. Instrum.,* **40**, 380 (1969)
234. G. Lawson and J.F.J. Todd, *Chem. Brit.,* **8**, 373 (1972)
235. A. Smekal, *Naturwissenschaften,* **11**, 873 (1923)
236. C.V. Raman and K.S. Krishnan, *Nature,* **121**, 501, 619 (1928)
237. B.P. Stoicheff, *Experimental Physics: Molecular Physics*, vol. 3 (ed. D. Williams), Academic Press, New York, p. 111 (1962)
238. S.P.S. Porto and D.L. Wood, *J. Opt. Soc. Am.,* **52**, 251 (1962)
239. H.E. Hallam, *Roy. Inst. Chem. Rev.,* **1**, 39 (1968)
240. A. Burawoy, *J. Chem. Soc.,* 1177 (1939)
241. K. Bowden and E.A. Braude, *J. Chem. Soc.,* 1068 (1952)
242. A.D. Buckingham, *Can. J. Chem.,* **38**, 300 (1960)
243. A. Ringbom, *Z. Anal. Chem.,* **115**, 332 (1939)
244. M.J.E. Golay, *Rev. Sci. Instrum.,* **29**, 313 (1958)
245. W.A. Anderson, *Rev. Sci. Instrum.,* **32**, 241 (1961)
246. E.V. Shpol'skii, *Soviet Phys. Usp. (Eng. trans.),* **3**, 372 (1960)
247. D. Farooq and G.F. Kirkbright, *Analyst,* **101**, 566 (1976)
248. E.M. Purcell, H.C. Torrey and R.V. Pound, *Phys. Rev.,* **69**, 37 (1946)
249. N.E.M. Bloembergen, E.M. Purcell and R.V. Pound, *Phys. Rev.,* **73**, 679 (1948)
250. A.J. Dempster, *Phys. Rev.,* **11**, 316 (1918)
251. 'Quantities, Units and Symbols', *A Report by the Symbols Committee of the Royal Society*, London (1975); *Addenda* (1981)
252. D.H. Whiffen (ed.), *Manual of Symbols and Terminology for Physicochemical Quantities and Units*, IUPAC, Pergamon Press, Oxford and New York (1979)
253. N.H. Davies, *Chem. Brit.,* **6**, 344 (1970); **7**, 331 (1971)
254. R.W. Engstrom, *J. Opt. Soc. Am.,* **37**, 420 (1947)
255. L. Dunkelman, W.B. Fowler and J. Hennes, *Appl. Opt.,* **1**, 695 (1962)

256. A. Cornu, *Advan. Mass Spectrometry*, **4**, 401 (1968)
257. C.A. Evans and G.H. Morrison, *Anal. Chem.*, **40**, 869, 2106 (1968)
258. R.H. Davis and A.S. Divatia, *Rev. Sci. Instrum.*, **25**, 1193 (1954)
259. R.G. Herb, R.H. Davis, A.S. Divatia and D. Saxon, *Phys. Rev.*, **89**, 897 (1953)
260. H.H. Willard, L.L. Merritt and J.A. Dean, *Instrumental Methods of Analysis*, 5th edn, Van Nostrand Reinhold, New York, p. 379 (1974)
261. E.L. Grove (ed.), *Analytical Emission Spectroscopy*, vol. 1, Dekker, New York, part 1, p. 32 (1971)
262. D.C. Hess, R.R. Marshall and H.C. Urey, *Science*, **126**, 1291 (1957)
263. L. Geiling, *Z. Angew Phys.*, **3**, 467 (1951)
264. P.A. Leighton and W.G. Leighton, *J. Phys. Chem.*, **36**, 1882 (1932)
265. D. Price, *Chem. Brit.*, **4**, 255 (1968)
266. D. Price and J.E. Williams (eds), *Time-of-Flight Mass Spectrometry*, Pergamon Press, Oxford (1969)
267. G.V.D. Tiers, *J. Phys. Chem.*, **62**, 1151 (1958)
268. 'Quantities, Units and Symbols', *A Report by the Symbols Committee of the Royal Society*, London (1975); *Addenda* (1981)
269. G. Milazzo and G. Cecchetti, *Appl. Spectrosco.*, **23**, 197 (1969)
270. J.T. Watson and K. Biemann, *Anal. Chem.*, **36**, 1135 (1964); **37**, 844 (1965)
271. L. Dorfman, *Chem. Rev.*, **53**, 47 (1953)
272. R.B. Woodward *et al.*, *J. Am. Chem. Soc.*, **63**, 1123, 2727 (1941); **64**, 72 (1942)
273. N.D. Lang, *Solid State Phys.*, **28**, 225 (1973)
274. R.M. Dagnell, K.C. Thompson and T.S. West, *Anal. Chim. Acta*, **36**, 269 (1966)
275. C. Veillon, J.M. Mansfield, M.L. Parsons and J.D. Winefordner, *Anal. Chem.*, **38**, 205 (1966)